MATERIALS SCIENCE AND TECHNOLOGIES

PHASE FORMATION AND SUPERCONDUCTIVITY IN COPPER OXIDE BASED YBCO AND RU-1212 AND RU-1222 SYSTEMS PREPARED BY SOL-GEL AND COPRECIPITATION TECHNIQUES

MATERIALS SCIENCE AND TECHNOLOGIES

Additional books in this series can be found on Nova's website under the Series tab.

Additional E-books in this series can be found on Nova's website under the E-book tab.

SUPERCONDUCTIVITY RESEARCH AND APPLICATIONS

Additional books in this series can be found on Nova's website under the Series tab.

Additional E-books in this series can be found on Nova's website under the E-book tab.

MATERIALS SCIENCE AND TECHNOLOGIES

PHASE FORMATION AND SUPERCONDUCTIVITY IN COPPER OXIDE BASED YBCO AND RU-1212 AND RU-1222 SYSTEMS PREPARED BY SOL-GEL AND COPRECIPITATION TECHNIQUES

YEOH LEE MOI

Nova Science Publishers, Inc.

New York

For permission to use material from this book please contact us:
Telephone 631-231-7269; Fax 631-231-8175
Web Site: http://www.novapublishers.com

NOTICE TO THE READER

The Publisher has taken reasonable care in the preparation of this book, but makes no expressed or implied warranty of any kind and assumes no responsibility for any errors or omissions. No liability is assumed for incidental or consequential damages in connection with or arising out of information contained in this book. The Publisher shall not be liable for any special, consequential, or exemplary damages resulting, in whole or in part, from the readers' use of, or reliance upon, this material. Any parts of this book based on government reports are so indicated and copyright is claimed for those parts to the extent applicable to compilations of such works.

Independent verification should be sought for any data, advice or recommendations contained in this book. In addition, no responsibility is assumed by the publisher for any injury and/or damage to persons or property arising from any methods, products, instructions, ideas or otherwise contained in this publication.

This publication is designed to provide accurate and authoritative information with regard to the subject matter covered herein. It is sold with the clear understanding that the Publisher is not engaged in rendering legal or any other professional services. If legal or any other expert assistance is required, the services of a competent person should be sought. FROM A DECLARATION OF PARTICIPANTS JOINTLY ADOPTED BY A COMMITTEE OF THE AMERICAN BAR ASSOCIATION AND A COMMITTEE OF PUBLISHERS.

Additional color graphics may be available in the e-book version of this book.

LIBRARY OF CONGRESS CATALOGING-IN-PUBLICATION DATA

Moi, Yeoh Lee.
  Phase formation and superconductivity in copper oxide based YBCO and
RU-1212 and RU-1222 systems prepared by sol-gel and coprecipitation
techniques / author, Yeoh Lee Moi.
      p. cm.
  Includes bibliographical references and index.
  ISBN 978-1-61122-504-4 (hardcover)
  1. High temperature superconductors. 2. Copper oxide superconductors. 3.
Yttrium alloys. 4. Ruthenium compounds. 5. Phase rule and equilibrium. I.
Title.
  QC611.98.H54M65 2010
  621.3'5--dc22
                              2010041594

Published by Nova Science Publishers, Inc. † New York

CONTENTS

PREFACE

Superconducting material $YBa_2Cu_3O_{7-\delta}$ (YBCO) had been synthesized via five wet chemistry techniques, there are sol-gel based on acetat-tartrate (ASG), citric gel (CT), coprecipitation (COP), sol-gel-solid state reaction (ASG-SSR) and coprecipitation-solid state reaction (COP-SSR). Infrared spectrum (FTIR) on all the sol-gel powders qualitatively showed the presence of OH and M-O group. The existence of -OH group has ability to increase the diffusion rate between metals during synthesis process. Thermal gravimetric analysis (TGA) for theses product show that there are four steps of mass loss. TGA curves show that the suitable calcinations temperature for ASG, CT, COP, and ASG-SSR are 620 °C, 600 °C , 500 °C and 450 °C respectively. Differential thermal analysis (DTA) for all the products showed a strong exothermic peak which is due to oxidation of organic compound in the sample. After calcinations, the products are synthesized at four different temperatures for four hour. The optimum temperature for YBCO synthesis prepared by ASG and CT are 920 °C and 900 °C respectively while the optimum synthesis temperature for COP, ASG-SSR and COP-SSR at 950 °C. The zero-resistance temperature ($T_{c\text{-zero}}$) for YBCO prepared by the five wet chemistry techniques are in the range of 83 K and 90 K. Additional of nano Ag on YBCO does not much change the $T_{c\text{-zero}}$ but enhanced the critical current density, J_c of the samples. YBCO with 10 wt% nano Ag showed the highest J_c of 2.2 A/cm^2. The presence of nano Ag in small amounts had the ability to pin the vortices effectively. XRD patterns showed that nano Ag addition in YBCO does not destroy the orthorhombic structure of the samples. $Y_{0.9}Ca_{0.1}Ba_{1.8}Sr_{0.2}Cu_3O_{7-\delta}$ sample is synthesized by ASG-SSR technique showed $T_{c\text{-zero}}$ 80 K. Three series of system ruthenium superconductor $RuSr_2GdCu_2O_{8-\delta}$ (Ru-1212), $RuSr_2(Gd,Ce)Cu_2O_{8-\delta}$ (Ru-1222) and RuSr$_{1.5}$Ca$_{0.5}$PbCu$_2$O$_{8-\delta}$ (RuPb-1212) had been prepared via ASG technique. System Ru-1212 is synthesized under oxygen flow has optimum parameter at 1030 °C for 24 hours showed $T_{c\text{-zero}}$ 45 K. System Ru-1222 with doping of Ce in Gd is synthesized under oxygen flow showed $T_{c\text{-zero}}$ 16 K and 30 K. Ratio Gd/Ce, $x=$ 0.5, 0.6 and 0.7 has been showed superconductivity. Superconductor system RuPb-1212 has synthesis temperature at 890 °C with $T_{c\text{-zero}}$ 20 K. The decreasing of synthesis temperature for the same formation in between RuPb-1212 and Ru-1222 is due to the lower melting point of Pb. The substitution of Pb^{2+} which has bigger size compared to Gd^{3+} had been increase the volume of tetragonal cell in RuPb-1212 compared to Ru-1212.

Chapter 1

INTRODUCTION

The discovery of superconductivity behavior in ceramic materials based on copper oxide in 1986 was impressive the world of science. This phenomenon against the traditional concept is the notion that ceramic is insulator materials. Superconductivity mechanism-based copper oxides is still a mysterious and not understood by scientists as a whole, but finding materials with higher critical temperatures were kept going on. This effort can be seen from the first discovery of superconductor in the critical temperature of mercury 4.2 K up to the highest critical temperature of 164 K in the copper oxide superconductors Hg-Ba-Ca-Cu-O in 1993 [Putilin et al. 1993; Schilling et al. 1993]

Nowadays, high-T_c superconductor (HTSC) has high demand value in term of application and commercials. Method of preparation is one important step in finding superconductors with higher critical temperature. Good preparation methods an addition strength in effort of discovery new formation HTSC with higher critical temperature. Solid-state reaction (SSR) method is the oldest, simplest and still most widely used method for preparing polycrystalline materials. In this process, high purity metal oxides or carbonates in powder form are mixed together and subjected to a series of heating and grinding at elevated temperatures. This method is not very sophisticated, nevertheless it is very effective; almost all the new high-T_c superconductor were first prepared by SSR. One of the major problems associated with this process is usually provides a sintered body which has porosity as high as 20-30% [Kalubarme et.al 2009]. It is presumed to be important to prepare dense superconducting ceramic with densities as close to the theoretical value as possible in order to improve the superconducting characteristics, particularly the critical current density, together with mechanical strength and stability against degradation by water and carbon dioxide.

Co-precipitation methods seem to be offered distinct advantages to SSR method. However, there are still some difficulties to be solved with these methods. For example, it is difficult to co-precipitate some of the metal ions from starting solution thoroughly. Furthermore, some of metal components tend to be lost in further wasting operations. In this context, the co-precipitation using hydrophilic organic solvent is attractive to avoid these disadvantages of precipitation process performed in aqueous solutions. A hydrophilic organic solvent molecule with strong affinity for water in aqueous solution takes water away from the otherwise hydrated cations in the aqueous solution [Tatsuya Yamamoto, 1988] .

Wet chemistry and particularly sol-gel processing is one of the most promising way to get and excellent homogenization at an atomic scale of the elements and high reactivity of the precursors. This chemical method is fast, cheap, reproducible and easy to scale up. By experimental results, we demonstrated that sol-gel processing is more efficient than solid-state reaction for obtaining superconductor phase. It is well known that the SGR method is superior to the SSR in term of chemical homogeneity and chemical reactivity, which are important requirement in obtaining optimal superconductor ceramics. Beside that, this wet chemistry reaction allows better control on particle size and morphology, loss of stoichiometry and avoids contamination of the sample. Moreover, the starting solution in the sol–gel method is a homogeneous mixture of the metal ions in the desired stoichiometry ratio. By using the SGR method, not only does the phase purity improves, the grain size is also decreases. However, this wet chemistry route cannot proceed if one or more metal components fail to dissolve in water or acid. Previously, Fujihara et al. (1992) reported that $YBa_2Cu_3O_{7-\delta}$ superconductor could be formed from reaction between $Y_2Cu_2O_3$ and $Ba_2Cu_3O_{5+\delta}$ which were produced during heat-treatment of the gel. $Ba_2Cu_3O_{5+\delta}$ which was successfully prepared by the conventional sol–gel method, was used as a precursor for Nd-123 preparation.

This chapter will discuss the background of the development of high temperature superconductors since the discovery of the first superconductor until now, the basic properties of high temperature superconductor, concepts and theories of conventional superconductors such as Barden-Cooper-Schrieffer (BCS) theory. In addition, this chapter will also describes various wet chemical methods in superconductors preparation. High temperature superconductor applications in the industry and the latest discoveries in a world of superconductors are also been touched.

1.2. BACKGROUND AND DEVELOPMENT ON HIGH CRITICAL TEMPERATURE SUPERCONDUCTOR

Superconductor is a special material that can allow an electric current through it without any electrical resistance. In addition, these materials also act as a perfect diamagnetic material when applied magnetic field medium [Abd-Shukor 2004]. Superconductivity phenomenon was first discovered by scientists Netherlands, Heiki Kamerlingh Onnes, of Leiden University in 1911, following his successful on produces liquid helium at 4.2 K temperature in 1908. After his discovery, several types of superconductors such as lead have been found at a temperature of 7.2 K, 8 K at the temperature of Niobium, Niobium nitrida the temperature 15 K Niobium and germanium at 23 K.

In 1981, copper oxide-based ceramic materials have been studied by a team of chemists from France [Er-Rakha et al. 1981]. Their studies only focus on the structure of material and not to the nature of electricity. In 1986, the nature of superconductivity was discovered by Bednorz and Müller at IBM, Zurich, Switzerland with the critical temperature of 35 K. In 1987, high-T_c superconductor materials $YBa_2Cu_3O_{7-\delta}$ (YBCO) has been found behave superconductivity at 92 K [Wu et al.1987]. This discovery of superconductor materials that have critical temperatures above boiling point of nitrogen, 78 K has started a new era in the development of superconductors field.

Bismuth compound (Bi-Sr-Ca-Cu-O or BSCCO) with the critical temperature of 110 K was discovered by Maeda et al. in 1988. This was followed by Sheng and Hermann, (1988) who reported Thallium compound (Tl-Ba-Ca-Cu-O or TBCCO) with the critical temperature of 125 K in the same year. In 1993, Schilling et al. and Putilin et al. discovered mercury compounds (Hg-Ba-Cu-O or HBCO) with the highest critical temperature up to now, with T_c 134 K. Critical temperature for this system can be increased to 164 K by applying pressure above 150 kbar [Chu et al. 1993].

Nagamatsu et al. (2001) reported that magnesium diborida (MgB_2) with the highest critical temperature of 39 K for conventional superconductors. In March 2008, Kamihara et al. report the discovery of superconductors $LaO_{0.89}F_{0.11}FeAs$ with the critical temperature of 26 K. This discovery is interesting because the material contains iron which have magnet properties but show superconductivity at high temperatures.

In the 20 years since the discovery of this ceramic superconductor materials, many high-T_c superconductors have been studied and synthesized. All types of superconductors whether conventional superconductors, organic, heavy Fermion and high-temperature superconductors, have scientific value from a different angle in the direction of understanding superconductivity. Table 1.1 shows the evolution in superconductors field according to the year of discovering.

Table 1.1. The important areas in superconductors development

Year	Scientist	Development
1911	Onnes	The discovery of the phenomenon of Superconductivity in mercury at a temperature of 4.2 K.
1933	Meissner & Ochsenfeld	The discovery of the nature of perfect diamagnetic (Meissner effect).
1934	London, M & London	Expect the pricking depth λ' with two fluid model.
1950	Ginzburg & Landau, L	Theory of superconductors proposed makroskopik.
1957	Abrikosov, A	Classify superconductors of Type I and II.
1957	Bardeen et al.	Produce microscopic theory (BCS Theory).
1962	Giaever, G & Josephson	To study the effects penerowongan quantized in superconductors.
1973	Gavalen	The discovery of Nb_3Ge with the critical temperature of 23 K.
1986	Bednorz, T. G & Müller, A	The discovery of La (Sr, Ba) CuO with the critical temperature of 30 K.
1987	Wu et al.	Discovery $YBA_2Cu_3O_{7-\delta}$ with the critical temperature of 90 K.
1988	Maeda et al.	Discovery of $Bi_2Sr_2Ca_2Cu_3O_{10}$ with the critical temperature 110 K.
1988	Sheng & Hermann	The discovery of Tl-Ba-Ca-Cu-O with the critical temperature 125 K.
1993	Putilin et al. Schilling et al.	The discovery of mercury-based superconductor system with copper oxide critical temperature 134 K.
1995	Bauernfeind et al.	The discovery of $RuSr_2LnCu_2O_8$ (Ru-1212) and $RuSr_2(Ln_{1+x}Ce_{1-x})Cu_2O_{10}$ (Ru-1222) with Ln = Sm, Eu and Gd.
2001	Nagamatsu et al.	The discovery of conventional superconductors MGB_2 with critical temperature 39 K.
2008	Kamihara et al.	The discovery of superconductors $Lao_{0.89}F_{0.11}FeAs$ with the critical temperature of 26 K.

1.3. WHAT IS A SUPERCONDUCTOR?

Superconductor is a metal, alloy, or compound that has two special properties because the main characteristics, namely:

I. Show zero resistance below critical temperature, T_c.

II. Have perfect diamagnetic properties when the magnetic field imposed is less than critical field, B_c.

When electric current flowing through the materials without any loss of energy that characterize as superconductivity material. Superconductor only show superconductivity nature when reach critical temperature, T_c. The critical temperature for one superconductor is depending on the type and nature of the material and also influenced by the purity and it homogeneous.

Superconductor is perfect diamagnetic materials, ie materials which can reject any imposed magnetic field provided the field is less than critical *field, B_c*. Magnetic field that exists in the superconductor material due to field imposed from outside can be canceled by the magnetic field produced by surface currents of superconductor materials. In other words, magnetic flux superconductor not allow to exist in the approach to all conditions, known as the Meissner effect. The detail on mechanism based high temperature superconductor copper oxides (types II superconductor) haven't get well explanations and understood by scientist world. Thus, the space of development and studies on superconducting materials are wide. As optimistic, the effort to finding material that can show superconductivity at room temperature still continued.

Superconductors can be divided into five classes according to natural characteristics such as its critical temperature, T_c, structure formula and mechanism of materials. There are conventional superconductors, heavy Fermion superconductors, organic superconductors, high temperature superconductors based on copper oxide superconductors and borokarbida-based and boronitrida. Table 1.2 show examples of five class superconductors and it critical temperature, T_c .

Table 1.2. Classes of superconductor

Class	Example	Critical temperature, T_c (K)
Conventional	Al, Pb, MgB_2 , Nb_3Sn Al, Pb, MgB_2, Nb_3Sn	≤ 39 K ≤ 39 K
Heavy Fermion	$CeCu_2Si$, $UBE_{13,}$ $CeAl_3$	< 1 K <1 K
Organic	$Rb_{2.7}Tl_{2.2}C_{60}$	< 43 K <43 K
Copper Oxide	$HgBa_2Ca_3Cu_4O_{11-\delta}$	< 134 K <134 K
Borokarbida and Boronitida	$LuNi_2B_2C$, $La_3Ni_2B_2N_3$	< 23 K <23 K

1.4. HISTORY OF DEVELOPMENTS IN WET CHEMISTRY METHOD PREPARATION ON SUPERCONDUCTOR

In solid-state reaction, the starting components are high purity metal oxides or carbonates which are quite inert, unreactive solids. Although they are well mixed at the level of individual particles (e.g. on a scale of 1 μm), on the atomic level they are very inhomogeneous. This method is intrinsically slow because the formation of the superconductivity phase proceeds via diffusion in the solid-state. Multiple grindings and prolonged thermal treatment at high temperature are necessary to obtain a single compound and enhance densification. Sometime, higher temperature and heat duration will bring about melting in some oxide or carbonate powders, thus homogeneity and stoichiometry of compound decrease. To achieve better mixing of the initial products, a lot of chemical preparation techniques for superconducting powder have been developed.

Sol-gel technique and the co precipitation are commonly used in the preparation of superconductor powders. This is because this technique not only involves chemicals that the cheaper metal salts such as metal acetate and metal nitrate, but it can also produce large quantities. Wet chemical technique is a method that is cheap, easy, and the result is expected because of good quality is high homogeneity and purity. Sol-gel method is actually not only limited to the sol-gel based tartrat-acetate, a variety of techniques based on wet chemical techniques such as gel polymerization and co-precipitation with a more commercially, cheap, fast, easy was designed by scientists to optimize the product of superconductors. This sort of effective techniques are very useful not only in commercially purpose, get additional strengths in discovery more high T_c superconductor material in advance.

Yamamoto et al. (1988) have successfully produced sub-micro oxide powder $Ba_2LnCu_3O_{7-\delta}$; Ln= Y, La, Ho, Dy superconductor by using sedimentation co-precipitation technique (COP). Oxalic Acid act as deposition agent while ethanol act as hydrophobic solvent which have strong affinity to water molecular. High purity precipitated power were annealed at 900 °C and form superconductor.

Palkar et al. (1991) found that the conventional method of solid-state reaction (SSR) in the preparation of $YBa_2Cu_4O_8$ (Y-124) is very complicated because its involving extremely high pressures and single phase is difficult to be obtained. They used conventional sol-gel method to solve the problem. Y-124 single phase with $T_{c\text{-zero}}$ value of 80 K had been synthesized at temperature 900 °C for 30 minutes at atmosphere pressure.

Baranauskas et al. (2001) have successfully synthesized $YBa_2(Cu_{1-x}Mn_x)_4O_8$ superconductor with acetate tratrate sol-gel method (ASG) under 1 atmosphere pressure oxygen flow. They used to modify pH of tartaric acid sol to pH 5.5 to prevent crystallization of copper acetate during gelation.

Halim et al. (1999) have made comparisons between the sol-gel method (SGR) with solid-state reaction (SSR) in $Bi_{2-x}Pb_xSr_2Ca_2Cu_3O_8$ (x=0.4, 0.5) superconductors preparation. XRD data obtained from SGR samples have 85% and 95% of 2223 phase which is higher percentage compared with sample from SSR has only show 75% and 78%. In addition, sample prepared by SGR method have higher critical temperature, $T_{c\text{-zero}} = 105$ °C compared with samples prepared by SSR, with $T_{c\text{-zero}} = 98$ °C.

Sin et al. (2000) applied alkyl amine gelation technique to produce (Hg, Re)-1223 superconductor powder which is very homogeneous with nano size. This Wet chemical method is fast, low-cost, safe and easily prepared.

Suryanarayanan et al. (2001) successfully synthesized $YBa_2Cu_3O_7$ by pyrolysis citrate gel method. Metal nitrates salt dissolved in hot distilled water. Citric acid, which has 50% moles over the total mole of metal ions was added. NH_4OH used to adjust and maintain the pH of the solution to pH 4.5. Then, ethylene glycol was added as chelating agent. Mixtures of solution was stirred by magnetic bar stirrer until it turn to viscosity gel. Gel is dried and ground into powder for further superconductors preparation.

Calleja et al. (2002) successfully using alkylamine sol-gel polymerization (AA) in preparation of superconductor powders. Superconductor phase Y-123, (Bi, Pb)-2212, and (Bi, Pb)-2223 were produced on a large scale in which 1 kg of each preparation.

Hamadneh et al. (2006) successfully synthesized high purity oxide powder with $Tl_{0.85}Cr_{0.15}Sr_2CaCu_2O_{7-\delta}$ superconductor system. In this study, they modified coprecipitation method based oxalate-2-isopropanol as hydrophobic solvent.

1.5. APPLICATIONS OF HIGH TEMPERATURE SUPERCONDUCTOR

Two special properties of superconductors, there are zero resistance and perfect diamagnetic will bring revolution to the power industry, transport and communications. With the discovery of high temperature superconductor materials based on copper oxide which show superconductivity and diamagnetic properties over liquid nitrogen boiling point has been extended for technology application and commercially use because the operating costs and refrigeration liquid nitrogen is low. While scientists are trying to understand the phenomenon and mechanism of superconductivity-based copper oxide, various tools and devices have been designed. Uses of superconductors can be divided into large-scale and small scale.

1.5.1. Uses Large Scale

Use equipment including large magnetic energy storage systems, magnetic resonance imaging (MRI), cable and transformer types of superconductors and MagLev (floating rail system prototype).

Perfect diamagnetic nature of superconductors has produced the inventions of floating train (MagLev). The train track is build up by high-T_c superconductor material while the wheel of train consist of strong magnetic material. When power is on, train track will generate diamagnetic flux to the train wheel, thus without any friction between the train and train track when move. MagLev trains can move at speeds exceeding 400 km/hour and as high as 10 cm from the floating platform. MagLev technology was used commercially in 1990 in Japan.

MRI (magnetic resonance imaging) was discovered in 1940s and used since year 1977. This system plays a vital role in the medical field because this equipment can detect soft tissue in the human body. By applying magnetic field generate from superconductors on patients body, hydrogen atoms that exist in bodily fluids and fat molecules can be focused to receive energy from the magnetic field. After that, energy will be released at a certain frequency that can be detected and displayed on the computer.

1.5.2. Small Scale Uses

Use equipment such as antennas, thin films, ray detection systems, Josephson junctions, fast oscilloscope and superconductor quantum interference device (SQUID) are small scale uses produce by the nature of zero resistance superconductor.

SQUID is sensitive flux detector which can detect small changes in magnetic flux. This sensitive equipment is very useful in the field of meteorology. Josephson junction is the switch that quickly oscilloscope used in fast operating at 10 GHz. In addition, superconductor chip that can transmit data quickly and effectively towards the development of a super-computer.

1.6. OBJECTIVE OF RESEARCH STUDY

Two systems of high temperature superconductor, ie YBCO (Y-123) and Ru-Cu-O (Ru-1212 and Ru-1222) has been synthesized and studied in terms of preparation techniques and electrical properties (transition critical temperature).

1.6.1. YBCO System

I) $YBa_2Cu_3O_{7-\delta}$ superconductor had been synthesized by using five types of wet chemical techniques, namely:

      a.   Sol-gel route based acetate-tartrat (ASG).
      b.   Sol-gel route based citrate gel complexion technique (CT).
      c.   Coprecipitation route (COP).
      d.   Sol-gel-solid state reaction route (SSR-ASG).
      e.   Coprecipitation- solid-state reaction route (SSR-COP).

These five types of wet chemical techniques use to prepare YBCO superconductor powders have been studied in terms of its method steps of preparation, temperature treatment (calcinations and annealing temperature), after that, the outcomes of YBCO superconductor samples are test and measure by FTIR, TGA /DTA, T_c, XRD and SEM for further analysis.

II) To study the effects of nano size silver (Ag) addition on $YBa_2Cu_3O_{7-\delta}$ by using ASG-SSR route. The product of YBCO-Ag superconductors were characterized in terms of critical temperature (T_c), critical current density (J_c), x-ray diffraction pattern (XRD), microstructures (SEM) and EDAX mapping.

III) Study on the doping effects of Ca into Y and Sr into Ba in YBCO system. Comparison was made between $Y_{0.9}Ba_{0.1}Ca_{1.8}Sr_{0.2}Cu_3O_{7-\delta}$ and $YBa_2Cu_3O_{7-\delta}$ in terms of critical temperature T_c, XRD pattern, microstructures (SEM) and EDAX.

1.6.2. Ru-Cu-O System

Three series systems of rutheno-cuprates superconductor had been synthesized via sol-gel-based acetate (ASG) route. FTIR characterization, T_c, XRD pattern and SEM were conducted on these samples for further analysis.

I) $RuSr_2GdCu_2O_{8-\delta}$ (Ru-1212) system was synthesized through ASG technique with three different annealing temperature, 950 ° C, 1000 ° C and 1030 ° C for 24 hours. The objective of this research is intended to stabilize the Ru-1212 phase by optimizing its calcinations temperature and duration of treatment under oxygen flow to produce samples that have the higher $T_{c\text{-zero}}$ value.

II) A series of $RuSr_2(Gd_{-x}Ce_x)Cu_2O_{10-\delta}$ (Ru-1222) system with, x = 0.4, 0.5, 0.6 and 0.7 have been synthesized through ASG technique with annealing temperature 1050°C for 24 hours. This research aims to investigate the changes in $T_{c\text{-onset}}$ and $T_{c\text{-zero}}$ value of the samples with increasing the x value.

III) New phase formation of $RuSr_{1.5}Ca_{0.5}PbCu_2O_{8-\delta}$ (RuPb-1212) system had been synthesized via ASG route with three different annealing temperature 850°C, 890°C and 920°C. The objective of this research is to stabilize the RuPb-1212 phase by optimizing its annealing temperature and duration of heat treatment under oxygen to produce samples with higher $T_{c\text{-zero}}$ value.

REFERENCES

Abd-Shukor, R. 2004. Introduction to Superconductivity in Metals, Alloys and Cuprates, Tanjung Malim: UPSI Publisher.

Baranauskas, A., Jasaitis, D., Kareiva, A., Haberkorn, R. & Beck, H.P. 2001. Sol-gel preparation and characterization of manganese-substituted superconducting $YBa_2(Cu_{1-x}Mn_x)_4O_8$ compounds. *Journal of the European Ceramic Society* 21 : 399-408.

Calleja, A., Casas, X., Serradilla, I.G., Segarra, M., Sin, A., Odier, P. & Espiell, F. 2002. Up-scaling of superconductor powders by the acrylamide polymerization method. *Physica C* 372-376 :1115-1118.

Er-Rakha, L., Micheall, C., Provost, J. & Raveaue, B. 1981. A series of Oxygen-deect perovskites containing Cu^{II} dan Cu^{III}: The oxides $La_{3-x}Ln_x$ $(Cu^{II}_{5-2y} Cu^{III}_{1+2y})O_{14+y}$. *Journal of Solid State Chemistry* 37: 151-156.

Fujihara, S., Zhuang, H.R., Yoko, T., Kozuka, H., Sakka, S., J.Mater.Res. 7 (1992) 2355.

Hamadneh, I., Yeong, W. K., Lee, T. H. & Abd-Shukor, R. 2006. Formation of $Tl_{0.85}Cr_{0.15}Sr_2CaCu_2O_7$ superconductor from ultrafine co-precipitated powders. *Materials Letters* 60: 734-736.

Halim, S. A., Khawaldeh, S. A., H, Mohamed. & Azhan, H. 1999. Superconducting Properties $Bi_{2-x}Pb_xSr_2Ca_2Cu_3O_y$ system derived via sol-gel and solid state routes. *Material Chemistry and Physics* 61: 251-259.

Nagamatsu, J., Nakagawa, N., Muranaka, T., Zenitani, Y. & Akimitsu, J. 2001. Superconductivity at 39 K in magnesium diboride. *Nature* 410: 63-64.

Kalubarme, R.S., Shirage, P.M., Iyo, A., Kadam, M.B., Sinha, B.B., Pawar, S.H. 2009. Synthesis and Magnetic Properties of $Bi_2Sr_2CaCu_2O_y$ superconductor by using Nitrate precursor. J. Superconductor Novel Magn 22: 827-831.

Kamihara, Y., Watanabe, T., Hirano, M. & Hosono, H. 2008. Iron-based layered superconductor $La[O_{1-x}F_x]FeAs$ (x = 0.05-0.12) with T_c = 26 K. *JACS communications* 130: 3296-3297.

Palkar, V. R., Guptasarma, P., Multani M. S. & Vijayaraghavan, R. 1991. Sol-gel: A novel method to synthesize $YBa_2Cu_4O_8$. *Physica C* 185-189: 479-480.

Putilin, S.N., Antipov, E. V. & Marezio, M. 1993. Superconductivity above 120 K in $HgBa_2CaCu_2O_{6+\delta}$. *Physica C*. Volume 212, Issue 3-4: 266-270.

Schiling, A., Cantoni, M., Cuo, J. D. & Ott, H. R. 1993. Superconductivity about 130 K in the Hg-Ba-Cu-O system. *Nature* 363: 56-58.

Sin, A., Odier, P., Weiss, F. & Nunez-Regueiro, M. 2000. Synthesis of (Hg,Re-1223) by sol-gel technique. *Physica C* 341-348 :2459-24600.

Suryanarayanan, R., Nagarajan, R., Selig, H. & Ben-Dor, L. 2001. Preparation by sol-gel, structure and superconductivity of pure and fluorinated LaBa$_2$Cu$_3$O$_{7-d}$. *Physica C* 361: 40-44.

Wu, M. K., Ashburn, J. R., Trong C. J., Hor, P. H., Meng, R. L., Gao, L., Huang, Z. J., Wang, Y. Q. & Chu, C. W. 1987. Superconductivity at 93 K in a new Mixed-phase Y-Ba-Cu-O compound systems at ambient pressure. *Physical Review Letter* 58: 908-910.

Yamamoto, T., Furusawa, T., Seto, H., Park, K., Hasegawa, T., Kishio, K., Kitazawa, K., Fueki, K. 1998. Processing and microstructure of highly dense Ba$_2$LnCu$_3$O$_7$ prepared from co-precipitated oxalate powder. *Superconducting Science and technology* 1: 153-159.

Chapter 2

BACKGROUND OF WET CHEMICAL TECHNIQUES PREPARATION ON HIGH TEMPERATURE CUPRATE OXIDE SUPERCONDUCTOR AND THEORIES OF SUPERCONUCTOR

INTRODUCTION

This chapter will focus on the study of characteristics, advantages and disadvantages of the five types wet chemical techniques used in high-T_c cuprate superconductor precursor preparation. In addition, basic characteristic properties of superconductor are well explained. Key features of high-T_c ruthenium and yttrium cuprate oxide superconductor system also suppressed in terms of unit cell structure, electrical and magnetic properties.

2.2. TECHNIQUES IN HIGH-T_c CUPRATE OXIDE SUPERCONDUCTOR POWDER PREPARATION

Since the discovery of high temperature cuprate oxide superconductor materials, various techniques on superconductor powder preparation have been studied to provide ultrafine, pure and high-quality oxide powders [Lee et al. 1994; Kakihana 1996; Mahesh et al. 1991; Medelius et al. 1989 and Van Bael et al . 1997]. Solid state reaction (SSR) is the most common technique used by scientists. High purify metal oxide and carbonate powder were mixed together and blended become homogeneous mixture powder. After that, the mixture is heated, cooled and pressed into pellets and then calcinate to a certain heat treatment. Formation of single phase superconductor through solid state reaction technique is dependent on the rate of cation-cation infiltration between metal oxide and carbonate during heat treatment. The kinetic of reaction rate in this technique is slow, therefore long calcination duration are needed. Equations $1/2Y_2O_3 + 2CuO + 3BaO = YBCO$ is dependent on grinding skill and heat treatment. The repetitive grinding is important to generate a homogeneous mixture. This technique is easy, simple and involves a sophisticated tool. However, this technique has several drawbacks which may be notified by poor sintering behavior, non-uniformity of particle size and shape, lack of reproducibility, multiphase character, and loss of stoichiometry due to volatilization of reactant at high temperatures. This necessitates repeated cycles of grinding and calcination and longer period of heating to complete the reaction [Halim et al. 1999]. Figure 2.1 shows the weakness in the solid state reaction process. In addition, high calcinations temperature and long heating duration will cause volatility of PbO, Tl_2O_3 and Bi_2O_3 [Rama Rao et al. 1996] as result destroy stochiometry of the superconductor compound.

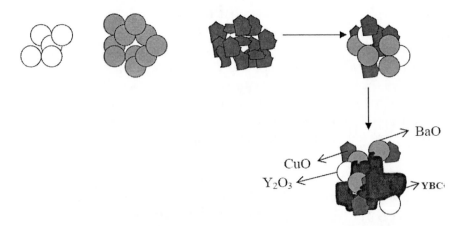

Figure 2.1. Solid state reaction between Y_2O_3, BaO and CuO with particle size of about 2-3 μm, the result of the product has largely raw materials that do not fully react.

Various wet chemistry techniques namely sol-gel technique and coprecipitation technique (COP) can overcome the weaknesses SSR technique. It is well known that wet chemical techniques may produce a more homogeneous cation compounds at the atomic scale. The sol-gel synthesis of oxide materials is a multistep process, which involves conversion of a precursor solution into sol or gel and subsequently transition of the gel to oxide compound. The chemical homogeneity with respect to distribution of cations in the gel often determines the compositional homogeneity of the final oxide powder. Sample powder produce by wet chemical technique requires low calcination temperature and short period of heating. The powder processing of high temperature ceramic superconductor is an important step for the fabrication of devices and characterizations of material properties. In the last decade, various wet chemistry techniques used to synthesis superconductor powder have been designed by scientists, there are sol-gel tartrat acetate or metal-chelate gel route [Baranaukas et al. 2001; Deptula et al. 1997; Palkar et al. 1991; Sin et al. 2000 and Van Bael et al.1999], sol-gel based citrate gel complexion or Pechini process [Peng et al. 1998], sol-gel based complexion *EDTA* [Calleja et al. 2002; Saitoh et al. 2002 and Tampieri et al. 2000], and coprecipitation technique [Hamadneh et al. 2006; Knaepen et al. 1997 and Xu et al. 1990, Yamamoto et al. 1988].

2.2.1. Metal-Chelate Gel Method
(Sol-Gel Based on Acetate-Tartarate)

The process of sol-gel-based acetate tartrate is similiar with metal-chelate gel reaction which involves hydrolysis and condensation reactions. The basic idea behind the metal-chelate gel formation method is to reduce the concentration of free-metal ions in the precursor solution by formation of soluble chelate complexes. From this point of view strong chelate agents such as tartaric acid or ethylene-diamine-tetraacetic acid (EDTA) are preferable used in synthesis of high-temperature superconducting oxides, which greately expands a range of experimental conditions such as pH of the solution, temperature, and metal concentrations where gelation can occur upon evaporation of the solvent. This method is a process of combination of various species single ion to a new ion compound (inorganic polymerization).

In high temperature superconducting YBCO preparaion, The hydrolysis of starting material metal salt $Y(CH_3COO)_3$, $Ba(CH_3COO)_2.xH_2O$, and $Cu(CH_3COO)_2.xH_2O$ will cause the cation of Y^{3+}, Ba^{2+} and Cu^{2+} solvated by water moleculars. This solvation during the hydrolysis process leads to the formation of a partial covalent bond between the metal ion and ligands, this can explain by 'partial charge model' developed by Livage et al (1988). When a metal salt dissolve in water, the cation of metal, M solvated by water moleculars $[M(H_2O)n]^{z+}$ show a strong tendency to release protons. The positive partial charge on the hydrogen atoms then increases and the water molecule, as a whole, become more acidic. Depanding on the magnitude of the electron transfer, the following reaction occur:

$$[M\text{-}OH_2]^{z+} \Leftrightarrow [M\text{-}OH_2]^{(z\text{-}1)+} + H^+ \Leftrightarrow [M\text{-}OH_2]^{(z\text{-}2)+} + 2H^+$$

Three kinds of ligands must then be considered in a non complexing aqueous medium:

a.   aquo ligands (OH_2)
b.   hydroxo ligands (-OH)
c.   oxo ligands (=O)

The partial charge model can be used in order to calculate the magnitude of electron transfer between ligands (oxo, hydroxo, aquo) and cation M^{z+}. Figure 2.2 showing charge-pH diagram illustrated the existence of these three possible water related ligands. Two important factors determining the degree of the hydrolysis include:

a.   the formal charge z of the cation M^{z+}
b.   the pH of the solution

Charge-pH diagram explain that low-valent cation (Z <+4) give aquo-hydroxo and/or hydroxo complexes over the whole range of pH, while high-valent cation (Z > +5), form oxo-complexes and/or oxo-hidrokso over the same range of pH. Tetravalent cation M^{4+}, Z = +4 are on the border line and therefore lead variety of ligands complexes will be formed, depending on the specific pH range.

In high T_c superconductor YBCO preparation, $Ba(CH_3COOH)_2$, $Y(CH_3COOH)_3$, and $Cu(CH_3COOH)_2$ are used as starting material under acidic condition, Since the cations Ba^{2+}, Y^{3+} and Cu^{2+} have valence lower than +4, then the result of hydrolysis reaction is aquo-hydroxo and/or complex hydroxo at the whole range of pH. Condensation reaction that occurs here is the process of hydroxo M-OH-M bridge formation between the two cations with acidic conditions (pH less than 7). Figure 2.3 shows the proposed sol-gel hydrolysis reactions in the formation of intramolecular chain -[Y-O-Ba-O-Cu-O]$_n$-. Olation process occurs corresponds to a nucleophilic substitution in which M-OH is the nucleophile and while H_2O group is the leaving group [Brinker & Scherer 1990 and Kakihana 1996].

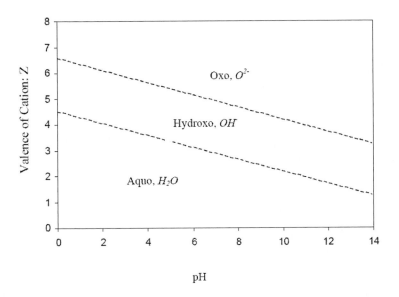

Figure 2.2. Graph showing pH charge-existence of three states of water in the form of oxo, hidroxo and aquo ligands [Brinker & Scherer 1990, Kakihana 1996].

Hydrolysis process between two mole of $Ba(CH_3COO)_2$ with water to form barium hydroxo complexes.

Barium hydroxo complex intramolecular

Hydrolysis reaction between one mole of $Y(CH_3COO)_3$ with water to form itrium hydroxo complexes.

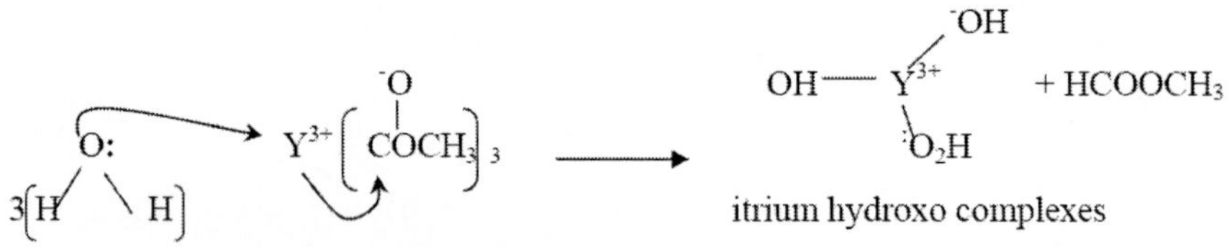

itrium hydroxo complexes

Hydrolysis reaction between three mol of $Cu(CH_3COO)_2$ with water to form cuprate hydroxo complexes.

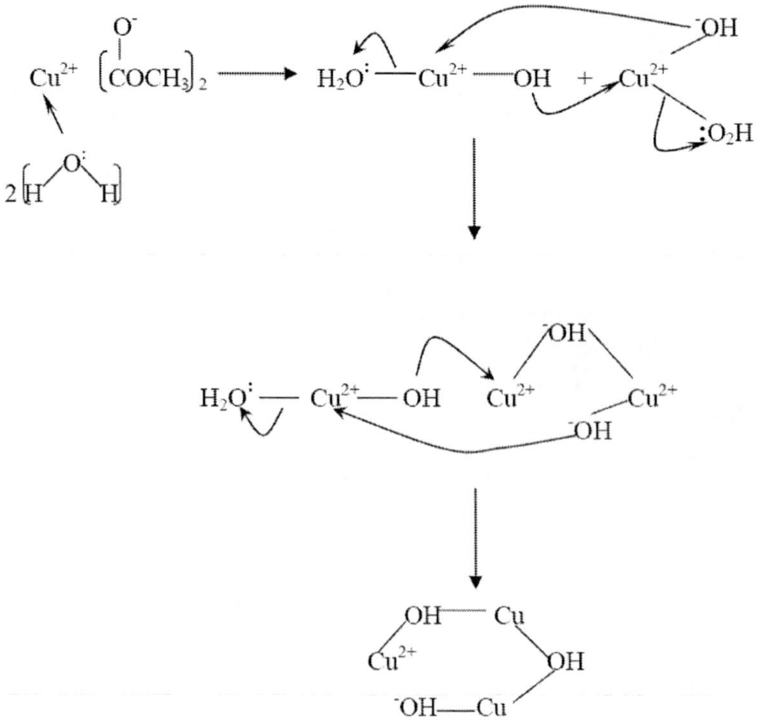

cuprate hydroxo complexes

When itrium, barium and copper hydroxo complexes meet together,olation (condensation) process occurs and cause the formation of inorganic polymeric complexes -[Y-O-Ba-O-Cu-O]$_n$-.

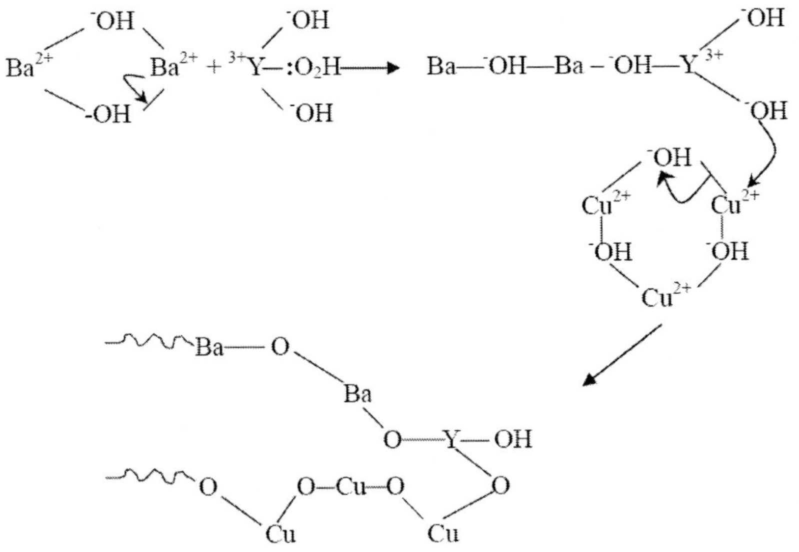

Inorganic polymeric oxide complexes -[Y-O-Ba-O-Cu-O]-$_n$

Figure 2.3. Reaction of hydrolysis-olation in acetate tartrate sol-gel process.

The existences of hydroxy groups (-OH) in metal hydroxo complexes have been provide additional oxygen content in the superconductor powder. This oxygen content plays an important role in efforts of optimize the charge carriers density in YBCO system. The product of homogeoneous inorganic polymeric oxide powder required low and shorten calcination temperature treatment for the formation of YBCO superconductor phase. This is due to the existence of Van der Waals bonds between the cation with the hydroxy groups results in the sol-gel reaction.

2.2.2 'In-Situ' Polymerization Method (Citric Gel Route)

The basic chemistry of this process is the dehydration reaction (esterfication) of a carboxylic acid and an alcohol. Immobilization of metal-chelate complexes in such a rigid organic polymer net can reduce segregation of particular metals during the decomposition process of the polymer at high temperatures. In this citric gel route process, the formation of a polymeric resin produced through polyesterification between metal chelate complexes by using citric acid (CA) and ethylene glycol (EG) [Kakihana 1996]. First of all, Y^{3+}, Ba^{2+} and Cu^{3+} cations form very stable chelate complexes with CA. Such metal-CA complexes formed can be further stabilized in EG as it processes two alchoholic hydroxyl fuctional groups with strong complexation affinities to metal ions. Successive ester reactions between CA and EG can occur to form a polyester resin, as CA contains three carboxylic acid groups (-COOH) in one CA molecule and EG contains two hydroxyl groups (-OH) in one EG molecule. The polyesterification reaction in this Pechini has been show in figure 2.4

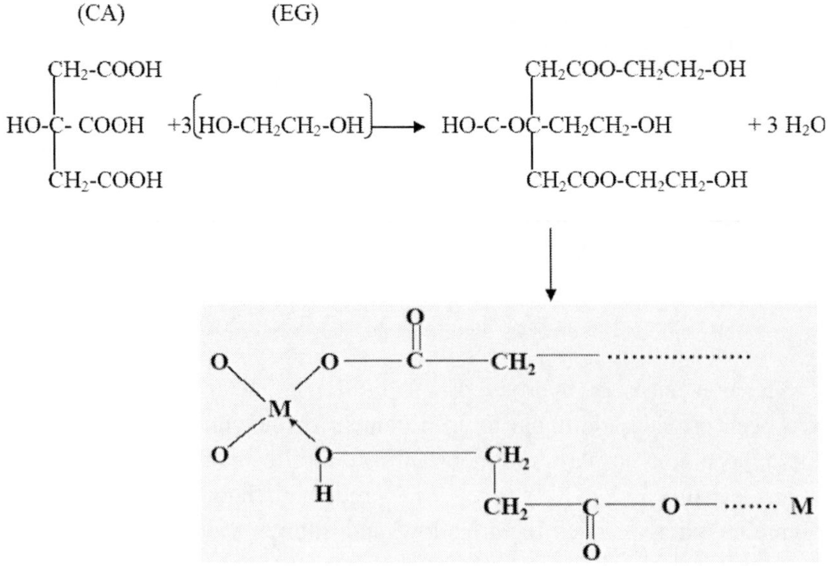

Figure 2.4. Pechini process, esterification between citric acid and ethylene glycol. [Kakihana 1996].

The principle of the Pechine method is thus to obtain a polymeric resin precursor comprising randomly branched polymer molecules throughout which the cations are uniformly distributed. Heating the of the polymeric resin at high temperatures (above 300°C causes a breakdown of the polymer. Eventhought the polymer are thermoplasticity, it is believed that less pronounced segregation of various cations would occour during the pyrolysis because of low cation mobility in such crowded branched polymers.

2.2.3. Coprecipitation Method

Coprecipitation method is also one of wet chemical techniques in the production of high quality superconducting oxides powder other than sol-gel method. The coprecipitation process involves the separation of solids containing various ionic species from aqueous solution phase. The main problem in this process is differences in solubility between the various precipitating phase strongly affect the precipitation kinetics of each metal ion component. That will resulting precipitate should be considered as heterogeneous mixture of fine particle, mean that the composition of each particle with respect to metal ions differing from one to another. Organic components that have multi-functional groups such as oxalates, tartarates and citrates are added to the solution as they can coordinate with more that one metal ion. The present of there precipitate agents can renders all the cations really insoluble in mother liquid. In addition, hydrophilic organic solvent molecular, such as ethanol that has strong affinity to H_2O in aqueous solution must be used to remove water from dissolving cation-cation complexes. This aims to encourage deposition reaction can occur with the more complete [Yamamoto et al. 1988]. Deposition with this technique can produce a very fine and purity heterogeneous mixture (100-500 nm). Solid powder precursor comprising of fine particles that greatly reduces the diffusion distances compared with those required for the solid state reaction, resulting in shorting reaction times and lower reaction temperature [Kakihana 1996].

2.2.4. Sol-Gel-Solid State Reaction Method

Sol-gel-solid-state reaction (SGR-SSR) and coprcipitation-solid state reaction (COP-SSR) are used when the problem of undisolveable for some of the metal oxide are happeded. Since Y_2O_3 cannot dissolve in any acidic solution and organic solvent, thus solid-state reaction carried out on -[O-Ba-Cu-O]$_n$- sol-gel powder with Y_2O_3 to form YBCO superconductors. Figure 2.5 shows the proposed of solid-state reaction between 1 mol of BaCuO with ½ Y_2O_3. This technique able to increase the degree of mixing between three components of cation Y, Ba, Cu. This condition is similar with the COP-SSR route, where the ultrafine powder of Ba_2Cu_3O compound are prepared through coprecipitation method. Powder of Ba_2Cu_3O compound mixed with Y_2O_3 at certain ratio to form the YBCO superconductor. However, difference calcinations temperature and duration are required for these multi-component oxides samples prepared through these five types of chemical methods to form a single phase of YBCO superconductors.

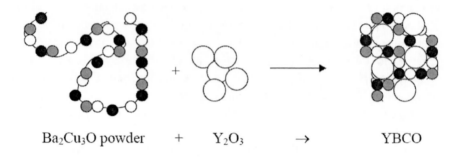

Ba_2Cu_3O powder   + Y_2O_3 $\rightarrow$ YBCO

Figure 2.5. Pattern proposed for solid stated reaction on $-[O-Ba-O-Cu]_n-$ sol-gel powder and $\frac{1}{2}$ Y_2O_3.

2.3. BASIC CHARACTER OF SUPERCONDUCTOR

Superconductor is a material that exhibits two special characteristic properties at certain condition. It show zero resistance and perfect diamagnetism, when it is cooled below its critical temperature, T_c and weak magnetic field, B are introduced.

2.3.1. Zero-Resistance

At room temperatures, superconductor materials such as pure metal, alloys, ceramic or organic material are usually not a good electrical conductor and even an insulator. At low temperatures, when reach its critical temperature, T_c, superconductor material can conduct electricity without any resistance. Meaning that the electric current flowing through it without any loss of energy. Figure 2.6 shows the electrical resistance against temperature for superconductors with T_c is the critical temperature.

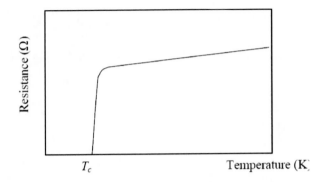

Figure 2.6. Graph of electrical resistance against temperature for the superconductor.

At higher temperatures it is a normal metal, and ordinarily is not a very good conductor. Copper, silver and gold which are much better conductor due to its purity of element, do not superconduct in any condition, while lead, tantalum, tin and mercury which are not a very good conductor and become superconductor at low temperature.

The phenomena of superconductivity mechanism superconductor can be explained by BCS theory proposed by Bardeen, Cooper and Schrieffer in 1957 since the discovery of superconductivity in 1911. Most of metals can conduct electrical current at room temperature but no show superconductivity behavior. When electrical current flows in the metal, at room temperature, electron conduction act as free moving electron scattered by the lattice vibrations thus resulting electrical resistance, as shown in figure 2.7 (a). When electric current flows in a metal, lead for example under its critical temperature, T_c superelectron consisting pairs of electrons are formed and flow in the material without any resistance. This condition is caused by electrons-phonons interaction which two electrons attract each other form Cooper pairs when superconduct.

According to the BCS theory as an electron passes through the lattice it distorts it since the metal ions are attracted to the passing electron. The distorted lattice draws in a second following electron because the positive charges are now more concentrated at the distortion and thus these two electrons travel together as a pair through the lattice, riding a wave of distortion, followed by other pairs, etc. The orderly travel of this "train" of electron pairs through the flexing lattice structure gives rise to the superconducting behavior, according to the BCS theory. In 1972 the three physicist were awarded the Nobel Prize for their work and theory.

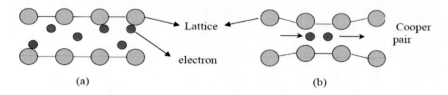

(a)　　　　　　　　　　　　　　　　　　　　　(b)

Figure 2.7. (a) distortion lattice cause vibrations due to electron flow normal condition of a conduction material (metal) and (b) shows the formation of Cooper pairs at low temperature [Charles et al. 1995].

2.3.2. Perfect Diamagnetic

Perfect diamagnetism is the second characteristic property of superconductor.

Superconductor materials show features perfect diamagnetic when medium megnet field is imposed in it under the critical temperature as observed in Figure 2.8. This means that it rejects the magnetic field imposed there on. This effect is called the Meissner effect-Ochsenfeld. Types I superconductors cancel all of magnetic flux imposed upon it when applied magnetic fields below the critical field B_c. Type II superconductor reject all magnetic flux when the magnetic field is applied is low. However, it allows partial penetrate pricking magnetic flux when the magnetic field applied is high. In areas of high magnetic field, Types II superconductor is not in perfect diamagnetic condition but in the mixed states.

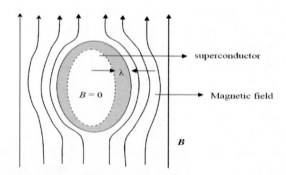

Figure 2.8. Fenomena diamagnet sempurna yang menunjukkan penolakan medanFigure 2.8. The phenomenon that shows the perfect diamagnetic field rejection magnet dari bahagian pendalaman superkonduktor apabila dikenakan interior of the superconductor magnet when medan magnet luar B . λ 'external magnetic field B are introduced. λ ' adalah kedalaman penusukan.is pricking depth.

Superconductors can be divided into two types, ie Type I and Type II. Types of superconductor are divided based on the transition behavior of material from superconduct state to normal state when the magnetic field imposed over critical field, B_c. Critical field is minimum strength of magnetic field needed to destroy it superconductivity for a superconductor material.

2.4. TYPES OF SUPERCONDUCTOR

2.4.1. Type I Superconductors

The behavior of Type I superconductor are more simple and straight forwards. It only exhibits perfect dimagnetism when applied magnetic fields below the critical field B_c, and become normal in higher applied magnetic field. At that time, electrical resistance will exist and be converted into zero diamagnetic moment. So, perfect diamagnetic nature will be lost when reach the critical field B_c.

Properties of Type I superconductors can be described in two aspects. The first is flux exclusion, way to state that the existence of surface currents, that reject the magnetic fields that penetrate into the material. When the magnetic field with flux density B_a is imposed on superconductor material, surface currents are surrounding the surface of the superconductor is found to exist. Surface currents will produce flux density B_i with the same magnitude but opposite direction to the imposed field B_a. Flux density produced will cancel the flux density imposed. Figure 2.9 shows the effect of the magnetic field on Type I superconductors.

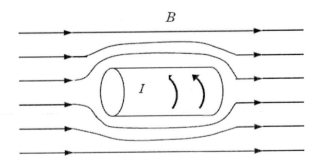

Figure 2.9. Effect of the magnetic fields on type I superconductors.

The second way assumes that the bulk diamagnetic field will cancel the imposed field and permeability of the material is zero. With permeability, μ_r is zero, thus the flux density in the interior materials superconductor is zero and this is a phenomenon called *Meissner effect* (flux expulsion). Figure 2.10 shows magnetization *(M)* against magnetic induction *(B)* for Type I superconductors. Figure 2.11 shows the scheme of magnetic induction *(B)* against temperature *(T)* for Type I superconductors.

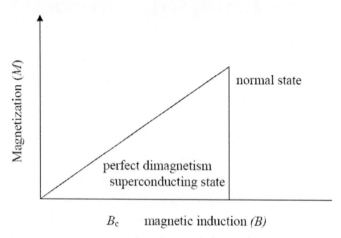

Figure 2.10. Magnetization *(M)* against magnetic induction *(B)* for Type I superconductor.

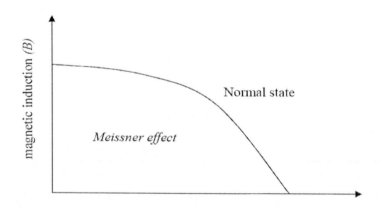

Figure 2.11. Scheme of magnetic induction *(B)* against temperature *(T)* for Type I superconductor.

2.4.2. Critical Current Density Type I superconductors

When a current I flowing through a wire radii r, a magnetic field will be generated from the surface current of wire as show as formula 2.1.

$$B_1 = \frac{\mu_0 I}{2\pi r} \tag{2.1}$$

B_1 found from the equation is directly proportional to the current I. If the magnetic field produced more than critical magnetic field, B_c, then the maximum current that can flow before superconductivity of the wire destroy are critical current, I_c.

If the magnetic field B transverse with $90°$ to the wire as shown in Figure 2.12, the transition to normal conditions on the surface will occur when the sum of the charged field and the field generated by currents equal to B_c. Then,

$$B_1 + 2B = B_c \tag{2.2}$$

and

$$B = B_c\text{-}2B = \frac{\mu_0 I}{2\pi r} \tag{2.3}$$

Critical current can be written as

$$I_c = \frac{2\pi r\,(B_c - B)}{\mu_0} \tag{2.4}$$

Critical current density (J_c) is the maximum amount of current per unit area of cross section area A, then

$$J_c = \frac{I_c}{A} \tag{2.5}$$

with a wire have radii r is πr^2. Replace the equation 2.4 to 2.5, obtained

$$J_c = \frac{2\,(B_c - 2B)}{\mu_0 r} \tag{2.6}$$

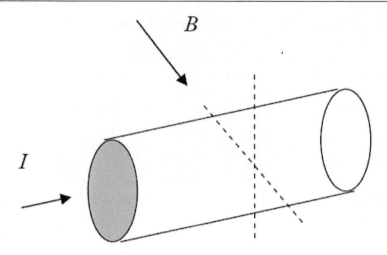

Figure 2.12. Wire lead current I imposed longitudinal field B.

2.4.3. Type II Superconductors

In 1957, Abrikosov has been classified into two types of superconductors. Since superconductor types II always consist of multicomponent compound, it seem like 'impurities' bring about the superconductivity properties. Anyway, he suggested that the presence foreign material does not cause the anomaly nature of superconducting properties but by it own nature superconductors. Therefore, he proposes Type II superconductors.

For Type II superconductors, the transitions from superconductivity state to normal state are slower and more complicated compared to type I superconductor. There are two types of critical field exist in this system, namely the critical field B_{c1} and the critical field B_{c2}. In the range of B_{c1} and B_{c2}, the magnetic flux imposed on the superconductor in the form of partial puncture of thin filament known as vortex. Vortex is the normal state that does not show superconductivity behavior. Figure 2.13 shows the partial pricking magnetic field for Type II superconductors. When stronger magnetic field is introduced to the superconductor, the density of vortex increase and the nature of perfect diamagnetic decrease as shown in Figure 2.14.

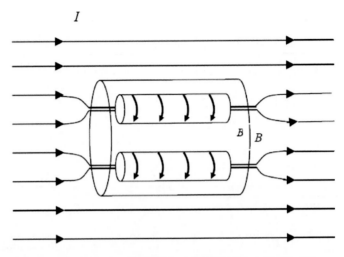

Figure 2.13. Pricking partial magnetic field for type II superconductors.

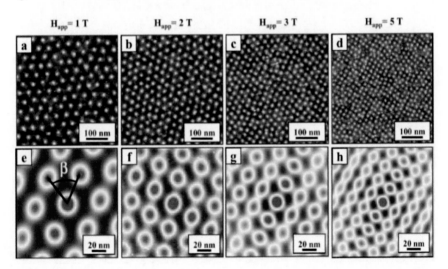

Figure 2.14. Vortex pattern (white spots) that exist in the superconductor (dark background) when introduced magnetic field 1 T, 2 T, 3 T and 5 T. Vortexs densities are directly proportional to the increase in magnetic field strength [Charles et al. 1995].

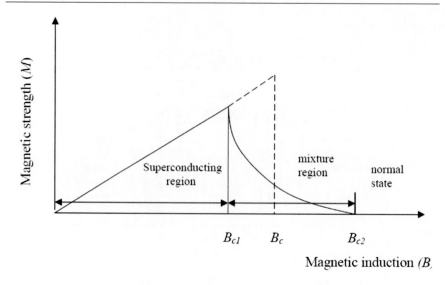

Figure 2.15. Magnetic strength *(M)* against magnetic induction *(B)* for Type II superconductors.

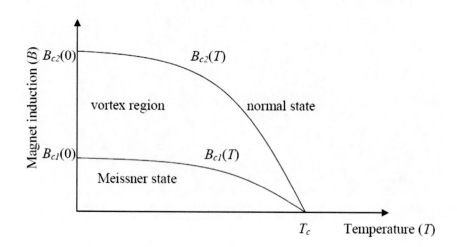

Figure 2.16. Schematic magnetic induction *(B)* against temperature *(T)* to Type II superconductors.

In the range of critical fields between B_{c1} and B_{c2}, superconductor in mixed region. At this stage, the magnetic field will exist in a superconductor, known as vortex. Superconductor material is converted into common conductor when the the strength of magnetic field applied excess the critical field B_{c2}. Figure 2.15 shows Magnetic strength *(M)* against magnetic induction

(B) for Type II superconductors. Figure 2.16 show the scheme of magnetic induction *(B)* against temperature *(T)* for Type II superconductors.

When the external magnetic field applied to the superconductor, the surface currents will exist and generate a magnetic field in the opposite direction with the external field. Currents that surround vortex are returned in the direction of the magnetic field parallel to the external field. Then flux trapped in vortex can be maintained because the interaction of surrounding current.

Interaction with the vortex current flow can use to determine critical current density, J_c of superconductor materials. Current flow will cause the vortex agitate and consume energy. Thus, it is important to prevent the agitations of vortex in Type II superconductors. Therefore, dopping of non-superconducting metal are included into the matrix of superconductor inorder to prevent the movement of vortex flux.

2.5. HIGH TEMPERATURE SUPERCONDUCTOR COPPER OXIDES

High temperature superconductor materials based on copper oxide (CuO_2) has a perovskite structure with oxygen defects play an important role during superconducting process. Figure 2.17 shows a layer of CuO_2 flanked by the ionic blocks. Beside $Nd_{2-x}Ce_xCuO_4$ system, block layer acts as electron receiver while the CuO_2 layer play role as hole conducting (proton receiver). Materials based on CuO_2 have a structure layer with a high anisotropy. Anisotropy in superconductor mean that, eventhought unit cell structure of superconductor material are in three dimension, conductivity behavior are greater in two dimensional CuO_2 layer compare to conductivity between the layers [Charles et al. 1995; Abdul Shukor 2004].

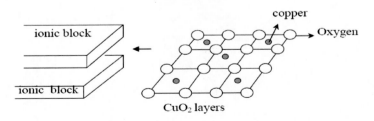

Figure 2.17. CuO_2 layer flanked by ions blocks and this is the important feature of high temperature copper oxide based materials.

One exclusive found in the copper oxide material is superconductivity properties of a substance that does not change significantly when the composition plane in the middle replaced [Rose-Innes and Rhoderick 1994]. This will provide opportunities for process of doping and replacement by other elements for optimizing a superconductor material and produce a new phase or system without affecting the properties owned by the superconductivity materials. For example, La_2CuO_4 is an insulator, but when part of the lanthanum atoms are replaced by strontium (doping) with the composition $La_{1.85}Sr_{0.15}CuO_4$, it becomes superconducting at critical temperature of 38 K.

2.6. HIGH TEMPERATURE COPPER OXIDE SUPERCONDUCTOR $YBa_2Cu_3O_{7-\Delta}$ (YBCO)

$YBa_2Cu_3O_{7-\delta}$ (YBCO) is a complex perovskite oxide material with oxygen defects plays an important role in superconductivity properties. Copper component in YBCO are in mixture state with present of both oxidation state Cu^{2+} and Cu^{3+} caused by oxygen defects. Cu^{2+} ions occupy layers of CuO_2 while Cu^{3+} ions occupy Cu-O chain. Superconductivity is the nature of YBCO is highly dependent on oxygen content. Material with the oxygen content δ <0.6 have orthorombic structure (Figure 2.18) with lattice parameters $a \neq b \neq c$ superconduct at the highest T_c, 92 K occurs when the oxygen content $O_{6.93}$. When the oxygen content material is reduced to δ> 0.6 the orthorhombic structure change becomes tetragon structure with lattice parameters $a = b \neq c$. This tetragon material shows insulator or semiconductor properties because oxygen occupy Cu-O chain has been taken out [Abd-Shukor 2004].

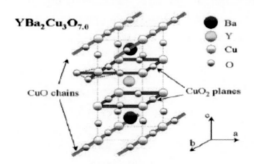

Figure 2.18. Orthorhombic structure of $YBa_2Cu_3O_{7-\delta}$ [Narlikar 2004].

2.7. EFFECT OF ARGENTUMS ADDITIONAL ON YBCO SUPERCONDUCTOR

YBCO is one of the applicable high-T_c superconductor material that is most studied. However, this system has several weaknesses, brittle, less elastic and weak currents transport properties. These poor mechanical properties of YBCO material can be overcome by the addition of metal Ag [Tepe et al. 2004]. Generally, it is believed that Ag diffuses into the grain boundary as a metal, thus improving the interconnections between the grains and enhances the electrical conductivity. It will cause the amount of weak links in the structure to decrease and improves the pinning capability [Salamati et al. 2001]. Addition of Ag was working as a foreign material to fix up the pin of vortex that exists in the mixed state, and current transportation properties are expected to increase. So far, not many results on the effect of nano size metallic Ag addition on $YBa_2Cu_3O_{7-\delta}$ have been reported. Many research has been conducted on addition of Ag metal into YBCO samples, such as the addition of micro-sized Ag metal [Salamati et al. 2001], Ag_2O powder [Abdelhadi et al. 1993], Ag_2O_2 powder [Sato et al. 2002] and $AgNO_3$ [Joo et al. 1999]. In this paper, we report on the synthesis of $YBa_2Cu_3O_{7-\delta}$–Ag through the sol-gel-solid-state method. The sol-gel method is generally employed to obtain chemical homogeneity and chemical reactivity.

One of the methods to enhance the pinning property of superconductor tapes is by introducing artificial pinning sites. Several techniques include heavy ion bombardment, proton irradiation, neutron irradiation and atomic substitutions, which have been used but may not be practical when applied in large-scale production. One of the solutions to overcome these problems is the addition of nanometer size particles as pinning centers. In this paper, we report on the effect of nano Ag addition on the transport current density and flux pinning enhancement in the YBCO superconductor prepared through the sol-gel method.

2.8. DOPPING EFFECT OF CALCIUM (CA) AND SRONTIUM (SR) IN YBCO SUPERCONDUCTOR

Many researchers have long attempted the substitution of Ca for Y in YBCO [Giri et al. 2005]. Ca-doped YBCO is particularly of interest in the study of the electronic properties of CuO_2 planes and the C-O chains. Ca^{2+} -

valence is smaller than the one of the substituted Y^{3+} that can result in increasing number of holes [Gasumyants et al. 2000; Lin et al. 2002]. Anyway, increasing Ca content to replace Y in Y-123 phase was found to be accompanied by decreasing oxygen content [4-11]. According to Awana et.al (1996) on replacement of Y by Ca, the oxygen content of $Y_{1-x}Ca_xBa_2Cu_3O_{7-y}$ (y $\approx$0.3) system remains nearly unchanged, till x =0.1 and drops sharply to x =0.15, which is followed by decreasing bond distances Cu-O planes with increasing Cu(2)-O(2)-Cu(2) angle. Oxygen content or hole concentration (n_h) of Ca-doped $Y_2Ba_4Cu_7O_{14-y}$ phase was found to decrease with increasing Ca^{2+} content and reduce the superconducting transition temperature as indicated by Chen et al. (1998). In contrast, Ca-doped $YBa_2Cu_4O_{8-y}$ phase was found that increase the T_c. The first Ca-doped Y-124 phase was reported by Miyatake et al (1989) with enhanced T_c to 91 K.

The studies on Sr-doped Ba site in Y-123 phase were conducted by several groups in the aspect of oxygen order-disorder explanation. Sr substitution for Ba is expected to suppress the oxygen mobility due to the contraction of the unit cell and consequent compression of the Cu(1)-type site [Veal et al. 1987], so T_c was reduced dramatically. Superconducting transition temperature, T_c for $Y(Ba_{1-x}Sr_x)_2Cu_3O_{7-y}$ with x= 0, 0.1, 0.2, 0.3, 0.4, and 0.6 were 91K, 85K, 83K, 80K and 78K, respectively [Ting et al.2001]. Bael et al. (1998) reported that Sr^{2+} substitutes preferably in the Ba site but not in Y site in $Y(Ba_{1-x}Sr_x)_2Cu_4O_{8-y}$ phase and the critical temperature are increased to 88K at 20% Sr substitution. So far, no work on the double doping of Ca and Sr in Y-123 phase was reported in the literature. In this present work, synthesis of $Y_{0.9}Ca_{0.1}Ba_{1.8}Sr_{0.2}Cu_3O_{7-\delta}$ phase was conducted by novel method, which combining the sol-gel and solid state reactions.

2.9. HIGH TEMPERATURE SUPERCONDUCTING CUPRATES BASED RUTHENIUM

The tripled-perovskite structure family represents one of the most important series of superconducting cuprates, which includes the widely studied $YBa_2Cu_3O_{7-\delta}$ system. Recently, member of this system, $RuSr_2GdCu_2O_{8-\delta}$ (Ru-1212) and $RuSr_2(Gd,Ce)Cu_2O_{8-\delta}$ (Ru-1222) has been the focus of new attention since the demonstration of the extremely unusual coexistence of ferromagnetism (FM) and superconductivity (SC) within the bulk material. Ginzburg theoretically argued that the exchange interaction responsible for FM leads to the break of the Cooper pair, thus destroy the superconductivity [Balchev et al 2005].

Both these FM dan SC phenomena interact through mutual exchange of intellectual and electromagnetic materials in the same compound [Balchev et al. 2005; Chu et al. 2000]. In 1950, Ginzburg and London has stated that superconductivity uniform order can not coexist with a uniform feromagnet order [Ginzburg & London 1956]. Matthias and Suhl (1958) have shown that the exchange field conditions tend to align the magnetic order in spin Cooper pairs with the oppressed nature of superconductivity [Matthias et al. 1958]. However, in 1959, Anderson and Suhl has shown that feromagnet order will not be uniform and also reduced by the existence of superconductivity.

In 1995, $RuSr_2LnCu_2O_{8-\delta}$, Ln = Sm, Eu, Gd, and who has the same structure with Nb(Ta)-1212 has shown superconductivity at critical temperature T_c ~40K [Bauernfeind et al. 1995]. Ru-1212 structure consists of a perovskite $SrRuO_3$ block layer, which connects the two CuO_2 planes in the shape of a pyramidal [Tang et al. 1997]. Ru is believed to appear in multivalensi state of Ru^{4+} and Ru^{5+} [Tallon et al. 2000; Tang et al. 1997]. Ru-1212 was also reported to contain domains of two crystallographically different superstructures [Yokosawa et al. 2004].

For $RuSr_2(Gd,Ce)Cu_2O_{8-\delta}$ (Ru-1222) system, in between two layers of CuO_2 plane is composed of three florite blocks, $(Gd_{1-x}Ce_x)_2O_2$ [Bernhard et al. 1999]. In both systems, Ru atoms which coordinate with a full octahedral of oxygen form the RuO_2 planes that replace the Cu–O chains which are present in the YBCO [Awana et al. 2002; Yokosawa et al. 2004]. The double layers CuO_2 are 'charge reservoir blocks' which are responsible for SC while RuO_2 layers are the origin of the FM properties [Hemberger et al. 2002; McLaughlin et al. 1999]. The electronic coupling between the magnetic and the superconducting layers which alternate on a (sub)nanometer scale shows that the RuO_2 and CuO_2 layers are separated by SrO layers which are likely to be insulating [Htrab et al. 2004; Nachtrab et.al 2004]. The crystal structure of hybrid ruthenocuprates, Ru-1212 and Ru-1222 are isostructural with the $YBa_2Cu_3O_{7-\delta}$ (Y-123) with Y, Ba, and Cu1 (the chain of copper atom) being completely replaced by Gd, Sr and Ru respectively for Ru-1212, (Gd,Ce), Sr, and Ru for Ru-1222 [Chmaissem et al. 2000]. Figure 2.19 shows the comparison between Ru-1212 structure with YBCO. Crystal structure of Ru-1212 and Ru-1222 are tetragon symmetry with space group P4/mmm and I4/mmm respectively as observed in Figure 2.20.

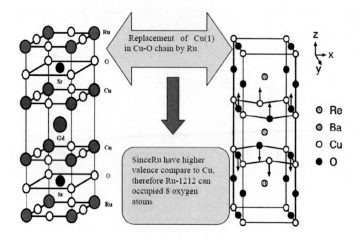

Figure 2.19. Comparison of crystal structures for Ru-1212 and Y-123

[Yokosawa et al. 2004].

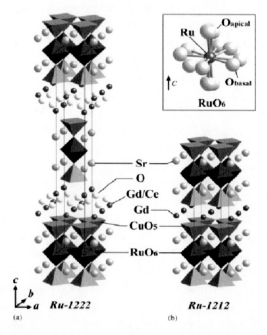

Figure 2.20. Comparison of crystal structure for Ru-1212 and Ru-1222 system. Ru is coordinated with a full octahedral of oxygen form the RuO_2 planes. [Yokosawa et al. 2004].

REFERENCES

Abdelhadi, M.M. & Zip, K.A. 1993. The behavior of the flux flow resistance in YBCO/(Ag$_2$O)$_x$. Superconductor Science and Technology 7: 99-102.

Abd-Shukor, R. 2004. Introduction to Superconductivity in Metals, Alloys and Cuprates, Tanjung Malim: UPSI Publisher.

Abrikosov, A. A., 1957. On the magnetic properties of superconductors of the second group. Sov. Physics. JETP 5: 1174-1182.

Awana, V. P. S. 2005. Magneto-superconductivity of rutheno-cuprates. A.V. Narlikar Edition, Frontiers in Magnetic Materials, Germany: p. 531. Springer-Verlag Publishing.

Awana, V. P. S., Ichihara, S., Karppinen, M. & Yamauchi, H. 2002. Comparison of magneto-superconductive properties of RuSr$_2$RuCu$_2$O$_{8-\delta}$ and RuSr$_2$Gd$_{1.5}$Ce$_{0.5}$Cu$_2$O$_{10-\delta}$. Physica C 378-381: 249-254.

Awana, V.P.S., Malik, S.K., Yelon, W.B. 1996. Physica C 262: 272.

Balchev, N., Kunev, B., Pirov, J., Mihova, G. & Nenko, K. 2005. Low temperature magnetoresistance in Ru-1222 superconductor. Materials Letters 59: 2357-2360.

Baranauskas, A., Jasaitis, D., Kareiva, A., Haberkorn, R. & Beck, H.P. 2001. Sol-gel preparation and characterization of manganese-substituted superconducting YBa$_2$(Cu$_{1-x}$Mn$_x$)$_4$O$_8$ compounds. *Journal of the European Ceramic Society* 21 : 399-408.

Bauernfeind, L., Widder, W. & Braun, H.F. 1995. Ruthenium-based layered cuprates RuSr$_2$LnCu$_2$O$_8$ and RuSr$_2$(Ln$_{+x}$Ce$_{1-x}$)Cu$_2$O$_{10}$ (Ln= Sm, Eu and Gd). *Physica C* 245:151-158.

Bernhard, C., Tallon, J. L., Niedermayer, C., Blasius, T., Golnik, A., Brucher, E., Kremer, R. K., Noakes, D. R., Stronach, C. E., Noakes, D. R., Stronach, C. E. & Ansaldo, E. J. 1999. Coexistence of ferromagnetism and superconductivity in the hybrid ruthenate-cuprate compound RuSr$_2$RuCu$_2$O$_8$ studied by moun spin rotation and dc magnetization. *Physical Review B* 59(21): 14099-14106.

Brinker, C.J. & Scherer. G.W. 1990. *Sol-Gel Science, The physics and chemistry of sol-gel processing.* New York: Academic Press.

Calleja, A., Casas, X., Serradilla, I.G., Segarra, M., Sin, A., Odier, P. & Espiell, F. 2002. Up-scaling of superconductor powders by the acrylamide polymerization method. *Physica C* 372-376 :1115-1118.

Charles, P. Poole., Horacio, Tr., Farach, A., Richard, J. & Creswick. 1995. *Superconductivity.* United States of America: Academic Press.

Chen, T. M., Yarng, S. L. & Lin, J. P. 1998. High-Pressure Synthesis and characterization of some Ca-Doped $R_2Ba_4Cu_7O_{15}$(R = Y, Sm, Dy) Cuprate Superconductors. *Chinese Journal of Physics* 36: 2-11

Chmaissem, O., Jorgensen, J. D., Shaked, H., Dollar, P. & Tallon, J. L. 2000. Crystal and magnetic structure of ferromagnetic superconducting $RuSr_2$-$GdCu_2O_8$. *Phys. Rev. B* 61:6401-6407.

Chu, C. W., Xue, Y. Y., Wang, Y. S., Heilman, A. K., Lorenz, R. L., Cmaidalka, J. & Dezaneti, L. M. 2000. The unusual phase-transitions in Ruthenate-Cuprate superconducting ferromagnets. *Journal of Superconductivity: Incorporating Novel Magnetism* 13(5): 679-686.

Deptula, A., Lada, W., Olczak, T., Goretta, K. C., Di-Bartolomea, A. & Casadio, S. 1997. Sol-gel process for the preparation of $YBa_2Cu_4O_8$ from acidic acetates/ammonia/ascorbic acid systems. *Pergamon, Material Research Bulletin* 32(3): 319-325.

Gasumyants,V.E., Elizarova, M.V., Vladimirskaya, E.V., Patrina, I.B. 2000. Physica C. 585: 341–348.

Ginzburg, V. L. & Landau, L. D. 1950. On the theory of superconductivity. *Zh. Eksp. Teor. Fiz* 20: 1064-1068.

Giri, R., Awana, A. P. S., Singh, H. K., Tiwari, R. S., Srivastava, O. N., Gupta, A., Kumaraswamy, B. V. & Kishan, H. 2005. Effect of Ca doping for Y on structural/ microstructural and superconducting properties of $YBa_2Cu_3O_{7-\delta}$. *Physica C* 419: 101-108.

Halim, S. A., Khawaldeh, S. A., H, Mohamed. & Azhan, H. 1999. Superconducting Properties $Bi_{2-x}Pb_xSr_2Ca_2Cu_3O_y$ system derived via sol-gel and solid state routes. *Material Chemistry and Physics* 61: 251-259.

Hamadneh, I., Yeong, W. K., Lee, T. H. & Abd-Shukor, R. 2006. Formation of $Tl_{0.85}Cr_{0.15}Sr_2CaCu_2O_7$ superconductor from ultrafine co-precipitated powders. *Materials Letters* 60: 734-736

Hemberger, J., Hassen, A., Krimmel, A., Mandal, P. & Loidl, A. 2002. Ferromagnetism and superconducting in pure and doped $RuSr_2GdCu_2O_8$. *Physica B* 212-213: 805-807.

Htrab, T. Nac., Koelle, D., Kleiner, R., Bernhard, C. & Lin. C. T. 2004. Intrinsic Josephson effects in the magnetic superconductor $RuSr_2GdCu_2O_8$. *Physical Review Letters* 92 :11.

Joo, J., Jung, S. B., Nah, W., Kim, J. Y. & Kim, T. S. 1999. Effects of silver additionals on the mechanical properties and resistance to thermal shock of $YBa_2Cu_3O_{7-\delta}$ superconductors. *Cryogenics* 39 :107-113.

Kakihana, M. 1996. Sol-gel preparation of high temperature superconducting oxides. *Journal of Sol-Gel Science and Technology* 6: 7-55.

Knaepen, E., Van Beal, M. K., Schildermans, I., Nouwen, R., D'Haen, J., D'Olieslaeger, M., Quaeyhaegens, C., Franco, D., Yperman, J., Mullens, J. & Van-Poucke, L. C. 1998. Preparation and characterization of coprecipitates and mechanical mixture of calcium-strontium oxalates using XRD, SEM-EDX and TG. *Thermochimica Acta* 318: 143-153

Lee, S., Shiyakhtin, O. A., Mun, M., Bae, M. & Lee, S. 1995. A freeze-drying approach to the preparation of $HgBa_2Ca_2Cu_3O_{8+x}$ superconductor. *Superconductor Science and Technology* 8: 60-64.

Lin, C.T., Liang, B., Chen, H.C. 2002. *J. Cryst. Growth* 778: 237–239.

Mahesh, R., Patvate, V. A., Parkash, O. & Rao, C. N. R. 1991. Investigations of the cuprate superconductors prepared by the combustion route. *Superconductor Science and Technology* 5: 174-179.

Matthias, B. T., Suhl, H. & Corennawit, E. 1958. Spin exchange in superconductors. *Physica Review Letter* 1:92-94.

McLaughlin, A.C., Zhou, W., Attfield, J. P., Fitch, A. N. & Tallon, J. L. 1999. Structure and microstructur of the ferromagnetic superconductor $RuSr_2GdCu_2O_8$. Physical Review B 60(10): 7512-7516.

Medelius, H. & Rowcliffe, D. J. 1989. Solution route to synthesize superconducting Oxides. Materials Science and Engineering A109: 289-292.

Meissner, W. & Ochsenfelf, R. 1933. Ein neuer effect bei eintritt der supraleitfahigkeit. Naturwissenschaften 21: 787-788.

Miyatake, T., Gotoh, S., Koshizuka, N. & Tanaka, S. 1989. T_{-c} increased to 90 K in $YBa_2Cu_4O_8$ by Ca doping. Nature 341: 41-44.

Nachtrab, T., Bernhard, C., Lin, C., Koelle, D. & Kleiner, R. 2006. The ruthenocuprates: natural superconductor- ferromagnet multilayers. Compres Rendus Physique 7: 68-85.

Narlikar, A. 2004. *Frontiers In Superconducting Materials.* 2004. Berlin Springer Verlag. Pg: 1-70.

Nenartavičienė, G., Petrėnas, T., Tautkus, S., Beganskienė, A., Jasaitis, D. & Kareiva, A. 2006. Characterization of strontium-substituted Y-124 superconducting compounds. *Lietuvos mokslų akademijos leidykla.* 17(4) :51–55.

Palkar, V. R., Guptasarma, P., Multani M. S. & Vijayaraghavan, R. 1991. Sol-gel: A novel method to synthesize $YBa_2Cu_4O_8$. *Physica C* 185-189: 479-480.

Peng, Z. S., Hua, Z. Q., Li, Y. N., Di, J. Ma, J., Chu, Y. M., Zhen, W. M., Yang, Y. L., Wang, H. J. & Zhao, Z. X. 1998. Synthesis and Properties of the Bi-Based Superconducting Powder Prepared by the Pechini Process. *Journal of Superconductivity* 11(6): 749-754.

Rama Rao, G. V., Varadaraju, U, V., Venkadesan, S. & Mannan, S. L. 1996. Synthesis of $(BiPb)_2Sr_2Ca_2Cu_3O_y$ Superconductors by sol-gel process. *Journal of Solid State Chemistry* 126: 55-64.

Rose-Innes, A. C. & Rhoderick, E. H. 1994. *Introduction to superconductivity*. Edition ke-2. Oxford: Pergamon Press.

Saitoh, H., Satoh, R,. Nakamura, A., Nambu, N. & Ohshio. S. 2002. Metal oxide powder synthesized with amorphous metal chelates. *Journal of Materials Science* 37: 4315-4319

Salamati, H., Babaei, A. A. & Safa, M. 2001. Investigation of weak links and the role of silver additional on YBCO superconductors. *Superconductor Science and Technology* 14: 816-819.

Sato, T., Nakane, H., Yamazaki, S. & Mori, N. 2002. Analysis of fluctuation conductivity in melt-textured $YBa_2Cu_3O_{7-\delta}$ superconductor with Ag-doping. *Physica C* 372-376: 1208-1211.

Sin, A., Odier, P., Weiss, F. & Nunez-Regueiro, M. 2000. Synthesis of (Hg,Re-1223) by sol-gel technique. *Physica C* 341-348 :2459-24600.

Tampieri, A., Gelotti, G., Lesca, S., Bezzi, G., La Torretta, T., M., G. & Magnani, G. 2000. Bi(Pb)-Sr-Ca-Cu-O (Bi-2223) superconductor prepared by improved sol-gel technique. *Journal of the European Ceramic Society* 20: 119-126.

Tang, K. B., Qian, Y. T., Yang. L. Y., Zhao, D. & Zhang, Y. H. 1997. Crystal structure of a new series of 1212 type type cuprate $RuSr_2LnCu_2O_z$. *Physica C* 282-287: 947-948.

Tang, K. B., Qian, Y. T., Zhao, Y. D,. Yang. L., Chen, Z. Y. & Zhang, Y. H. 1996. Synthesis and characterization of a new layered superconducting cuprate: $RuSr_2(Ce,Gd)_2Cu_2O_z$. *Physica C* 259: 168-172.

Tepe, M., Avci, I., Kocoglu, H. & Abukay, D. 2004. Investigation of in weak-link profile of $YBa_2Cu_{3-x}Ag_xO_{7-d}$ superconductors by Ag doping concentration. *Solid State Communications* 131: 319-323.

Van Bael, M.K., Kareiva, A.¨, Nouwen, R., Schildermans, I., Vanhoyland, G., Haen ,J., D'Olieslaeger, M., Franco, D., Mullens, J. & Poucke, L. C. Van. 1999. Sol-gel synthesis and properties of $YBa_2(Cu_{1-x}M_x)_4O_y$ (M=Co, Ni) and effects of addition replacement of yttrium by calcium. *International Journal of Inorganic Materials* 1: 259-268

Van Bael, M.K., Knaepen, E., Kareiva, A., Nouwen, R., Schildermans, I., Vanhoyland, G., D'Haen, J., D'Olieslaeger, M., Franco, D., Mullens, J. &

Poucke, L. C. Van 1999. Study of different chemical methods to prepare ceramic high-tempature superconductors. *Physica C* 98: 82-87.

Van Bael, M.K., Kareiva, A., Vanhoyland, G., Olieslaeger, J.D'., Franco, D., Quaeyhaegens, C., Yperman, J., Mullens, J. & Poucke, L. C. Van. 1998. Enhancement of T_c by substituting strontium for barium in the $YBa_2Cu_4O_8$ superconductor prepared by a sol-gel method. *Physica C* 307: 209-220

Xu, Q., Cheng, T., Li, X., Fan, C., Wang, H., Mao, Z., Peng, D., Chen, Z. & Zhang, Y. 1990. Study on Bi(Pb, Sb)SrCaCuO superconductors prepared by the oxalate co-precipitation technique. *Superconductor Science and Technology* 3: 373-376.

Yamamoto, T., Furusawa, T., Seto, H., Park, K., Hasgawa, T., Kishio, K., Kitazawa, K. & Fueki, K. 1988. Processing and microstructure of high dense $Ba_2LnCu_3O_7$ prepared from co-precipitated oxalate powder. *Superconductor Science and Technology* 1: 153-159.

Yokosawa, T., Awana, V. P. S., Kimoto, K., Takayama-Muromachi, E., Maarit, K., Yamauchi, H. & Matsui, Y. 2004. Electron microscope studies of nano-domain structures in Ru-based magneto-superconductors: $RuSr_2Gd_{1.5}Ce_{0.5}Cu_2O_{10-\delta}$ (Ru-1222) and $RuSr_2GdCu_2O_8$ (Ru-1212). Ultramicroscopy 98: 283-295.

Chapter 3

EXPERIMENTAL TECHNIQUES

INTRODUCTION

There are a number of ways that ceramic powders are synthesized today, including, but not limited to, solid-solid, solid-gas, and solution processes. The most common of these are solution processes [Sarun et al. 2006]. There are numerous methods of producing powders from solutions including co-precipitation, spray drying, freeze-drying, sol gel, combustion synthesis, and others [Kalubarme et al. 2009]. Recent study was done by Chen et al. (2010) success to synthesis high-performance YBCO superconducting films by fluorine-free YBCO sol with high $T_{c\text{-zero}}$ (> 94 K) and J_c (>10^6 A/cm^2). The ultimate goal that all of these processes have, are to produce cheap, uniformly sized, highly pure powders that have very fine particle size and little agglomeration [Makan Chen et al. 2004]. This section describes five wet chemistry of techniques used in the high temperature superconducting copper oxide compounds preparation and the test experiments used to study the different properties of the samples. The synthesized sol-gel powder samples were characterized by infrared spectroscopy using a Perkin-Elmer FTIR spectrum (FTIR). The samples were mixed with dried KBr and pressed into pellets. Thermo gravimetric analysis (TGA) and differential thermal analysis (DTA) on sol-gel powder were obtained on a Perkin-Elmer Model 1605. Characterization of the superconducting sample by standard four-probe technique was used to measure the temperature depandence of resistivity in the range 20-300 K is performed in the chamber cooling system. The introduction of phase and lattice parameters with X-ray diffraction (XRD) carried out further on superconducting compounds. In summary, five wet chemical techniques used in the preparation of superconducting oxide powder are sol-gel citric (CT), sol-gel method based on acetate-tartrat (ASG), coprecipation method (COP), coprecipitation-solid stated reaction method (COP-SSR), and sol-gel-solid state reaction (ASR-SSR).

In this study, the superconducting $YBa_2Cu_3O_{7-\delta}$ (YBCO) sample has been prepared through the five wet chemical methods and the comparisons of quality samples in terms of electrical resistance changes with temperature and X-ray diffraction have been done. A series of $YBa_2Cu_3O_{7-\delta}$ + n% nano-Ag, a total of five samples synthesized and studied by the method of ASG-SSR. Ca and Sr doping in the YBCO system has synthesized via ASG method. After that, the comparison in term of T_c value, phase formation by XRD pattern, microstructure by SEM have been studied between the new system of $Y_{0.9}Ca_{0.1}Ba_{1.8}Sr_{0.2}Cu_3O_{7-\delta}$ with YBCO sample. Ruthenium system in Ru-1212 phase, $RuSr_{1.5}Ca_{0.5}PbCu_2O_{8-\delta}$ and $RuSr_2GdCu_2O_{8-\delta}$ are synthesized by the ASG method. Ru-1222 phase with series $RuSr_2(Gd_{2-x}Ce_x)Cu_2O_{8-\delta}$ (x=0.4, 0.5, 0.6, 0.7, 0.8) synthesized by the method of ASG. The diagram below is a summary of the scope of this project.

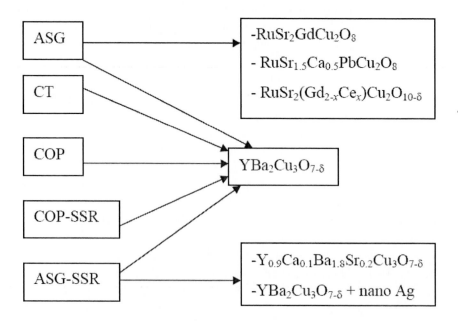

Figure 3.1. Summary of research scopes.

3.2. TECHNIQUES OF YBCO
SUPERCONDUCTOR PREPARATION

3.2.1. Citric Gel Route (CT) of YBCO
Superconductor Preparation

Acetate salt and nitrate salt which capable dissolve in weak acid are used as initial raw material. $Y(CH_3COO)_3.H_2O$, $Ba(CH_3COO)_2.H_2O$, and $Cu(CH_3COO)_2.H_2O$ dissolved in 0.05 M acetic acid solution with stoichiometric ratio. The mixture of solution were stirred with a magnetic bar and for one hour under 80 °C. An aqueous solution of ammonium hydroxide (NH_4OH) was added to the reaction mixture to adjust the pH to remain at pH 4.5. Citric acid (CA) with 0.45 ratio of the number of moles of copper acetate dissolved in 0.25 mol of ethylene glycol (EG). The mixture of CA/EG solution were then added to the blue reaction mixture solution [Suryanarayanan et al. 2001]. This is also define as polimerization method, CA and EG as the monomer for forming the polymeric matrix. The blue mixture solution were stirred for one hour to ensure that all the components were fully dissolve and then placed in an oven set at 100 °C. After 24 hours, the solution became dark blue glassy gel. Gel ground into fine-grained blue powder and tested with an infrared spectrometer (FTIR). The precursor powder was calcined in a furnace at a temperature of 800 °C for 12 hours to remove all organic materials. After that, the black oxide powder were reground in an agate mortar to upgrade it homogeneity. The powders were palletized with a diameter of 13 mm and 2 mm thick by hydraulic press at a pressure of 6-8 tons SPECAC. Annealed process were carried out in 2408-Carbilite tube furnace on these pellets form superconductor oxide at temperatures of at 880 °C, 900 °C, 920 °C and 950 °C respectively for 4 hours. Figure 3.2 shows the citric gel route technique used in the preparation of YBCO superconductors.

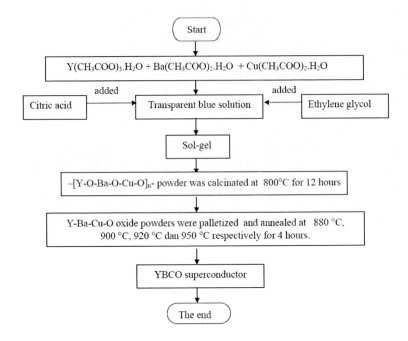

Figure 3.2. Citric gel technique preparation of YBCO superconductor.

3.2.2. Acetate-Tartrate Sol-Gel Route (Metal-chelate gel method) of YBCO Superconductor Preparation

As starting compounds, stoichiometric amounts of analytical grade $Y(CH_3COO)_3.H_2O$, $Ba(CH_3COO)_2.H_2O$, and $Cu(CH_3COO)_2.H_2O$ dissolved in 0.05 M acetic acid solution. The mixture of solution were stirred with a magnetic bar and for one hour under 80 °C. An aqueous solution of ammonium hydroxide (NH_4OH) was added to the reaction mixture to adjust the pH to remain at pH 4.5. An aqueous solution of tartaric acid ($C_4H_6O_6$) with a molar ratio to copper fixed at 0.45 was added and the reaction solution was stirred for 1 hour. Tartaric acid acts as a chelating agent to complex the copper ion before gelation [Zalga et al. 2006]. To prevent the problem of crystallization copper acetate during gelation, a dilute ammonium hydroxide, NH_4OH was added to adjust the solution to pH 5.5 [Yeoh et al. 2008]. In order to complete the hydroxylation and olation in the sol-gel process, the blue transparent solution was stirred for several hours in an open beaker. After that, the blue transparent solution was kept in to the oven at 100 °C for 24 hours then it will turn to glassy blue gel. Gel ground into fine-grained blue powder and tested with an infrared spectrometer (FTIR). The precursor powder was calcined in a furnace at a temperature of 800 °C for 12 hours to remove all organic materials. After that, the black oxide powders were reground in an agate mortar to upgrade it homogeneity. The black powders were palletized with a diameter of 13 mm and 2 mm thick by hydraulic press at a pressure of 6-8 tons SPECAC. Annealed process were carried out in 2408-Carbilite tube furnace on these pellets form superconductor oxide at temperatures of at 880 °C, 900 °C, 920 °C and 950 °C respectively for four hours. Figure 3.3 shows the acetate-tartrate sol-gel route used in the preparation of YBCO superconductors.

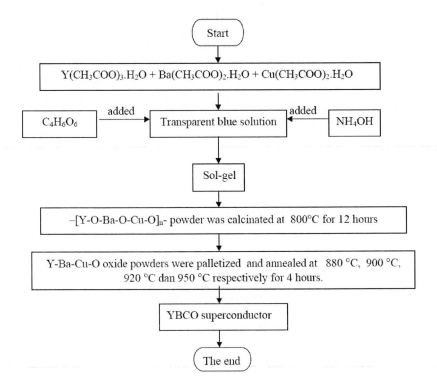

Figure 3.3. Acetate-tartrate sol-gel route preparation of YBCO superconductor.

3.2.3. Coprecipitation Route of YBCO Superconductor Preparation

Metal salts with stoichiometric amount of $Y(CH_3COO)_3.H_2O$, $Ba(CH_3COO)_2.H_2O$, and $Cu(CH_3COO)_2.H_2O$ with a purity of 99 % are the starting materials for metal solution preparation. All the chemicals were dissolved in dilute acetic acid 0.02 M with intermediate stirring for several hours at 70 °C to form transparent blue solution. Oxalic acid was dissolved in ethanol to have a concentration of 0.5 M. Blue metal solution was added to the stirred oxalic/ethanol solution under ice bath or 0 °C. The transparent blue solutions will start precipitate immediately and a uniform stable, blue suspension was obtained. The slurry was obtained after filtration and drying process, then follow by calcinations in box furnace at 880 °C for 12 hours. The black powders were palletized with a diameter of 13 mm and 2 mm thick by hydraulic press at a pressure of 6-8 tons SPECAC. Annealed process were carried out in 2408-Carbilite tube furnace on these pellets form superconductor oxide at temperatures of at 880 °C, 900 °C, 920 °C and 950 °C respectively for 4 hours. Figure 3.4 shows the acetate-tartrate sol-gel route used in the preparation of YBCO superconductors.

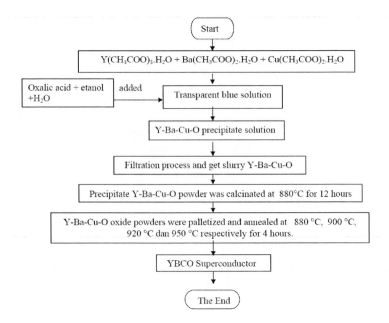

Figure 3.4. Acetate-tartrate sol-gel route preparation of YBCO superconductor.

3.2.4. Sol-Gel-Solid State Reaction (ASG-SSR) of YBCO Superconductor Preparation

Since yttrium oxide, Y_2O_3 unable dissolved in water and acid solvent, hence solid-state reaction was carried out on -$[Ba-O-Cu-O]_n$- gel powder and Y_2O_3 to form YBCO superconductor. -$[Ba-O-Cu-O]_n$- gel powder were prepared by acetate-tartarte sol-gel route which is same as method in figure 3.3. One mole of -$[Ba-O-Cu-O]_n$- and 0.5 mole of Y_2O_3 were physically mix and ground with mortar for 1 hour to upgrade its homogeneity. Then calcinations was done on the homogeneous powder in box furnace at 800 °C for 12 hours in order to eliminate the organic residue in sol-gel powder and provide initial reaction to the mixture. The powders were palletized with standard diameter of 13 mm and 2 mm thick by hydraulic press at a pressure of 6-8 tons SPECAC. Annealed process were carried out in 2408-Carbilite tube furnace on these pellets form superconductor oxide at temperatures of at 880 °C, 900 °C, 920 °C and 950 °C respectively for 4 hours. Figure 3.5 shows the sol-gel-solid state reaction route used in the preparation of YBCO superconductors.

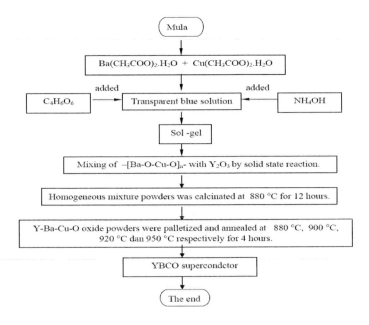

Figure 3.5. Sol-gel-solid state reaction route used in the preparation of YBCO superconductors.

3.2.5. Coprecipitation-Solid State Reaction (COP-SSR) of YBCO Superconductor Preparation

This technique is similar case with ASG-SSR which is involve solid-state reaction on BaCuO precursor and Y_2O_3. In this case, BaCuO precursor powder was prepared coprecipitation route which is same as the method as shown in figure 3.4. Figure 3.6 shows the summary of coprecipitation-solid state reaction (COP-SSR) route used in the preparation of YBCO superconductors.

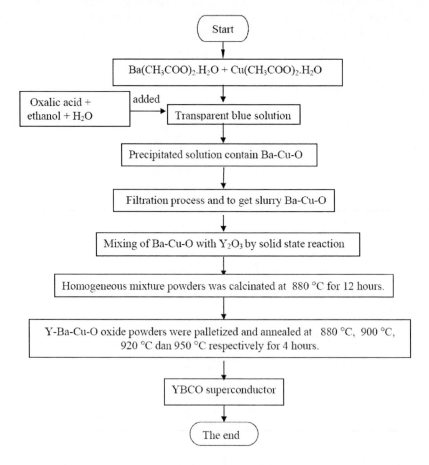

Figure 3.6. Coprecipite-solid state reaction route used in the preparation of YBCO superconductors.

3.2.6. Preparation of YBCO Oxide + n% Nano Argentum Powders by ASG-SSR Route

Certain percentage of nano size of argentum was added to the dried powder with nominal composition $YBa_2Cu_3O_{7-\delta}$. Solid state reaction was carried out on one mole of BaCuO powder with 0.5 mole of Y_2O_3 plus with n percentage of nano argentum. BaCuO powder was prepared by arsetate-tartarate sol-gel technique which is show as figure 3.4.

$$0.5\ Y_2O_3 + \text{-}[Ba\text{-}O\text{-}Cu\text{-}O]_n\text{-} + n\ \%\ nano\ Ag = YBCO + n\%\ Ag$$

where $n = 0, 2, 5, 10, 15$

The homogeneous solid state mixture of YBCO+ n % Ag were calcinated in box furnace at 880°C for 12 hours in order to eliminate the organic residue in sol-gel powder and provide initial reaction to the mixture. The powders were palletized with standard diameter of 13 mm and 2 mm thick by hydraulic press at a pressure of 6-8 tons SPECAC. Annealed process were carried out in 2408-Carbilite tube furnace on these pellets form superconductor oxide at temperatures of at 880 °C respectively for 18 hours to make a final bond forming reaction.

3.2.7. Preparation of $Y_{0.9}Ba_{1.8}Sr_{0.2}Cu_3O_{7-\delta}$ by ASG-SSR Route

Solid state reaction was carried out on one mole of Ca-Ba-Sr-Cu-O sol-gel powder plus 0.45 mole of Y_2O_3.

$0.45\ Y_2O_3 + Ca_{0.1}Ba_{1.8}Sr_{0.2}Cu_3O = YCaBaSrCuO = Y_{0.9}Ca_{0.1}Ba_{1.8}Sr_{0.2}Cu_3O_{7-\delta}$

Technique use to prepare Ca-Ba-Sr-Cu-O sol-gel powder was similar with figure 3.4. Starting material use in this project are $Ca(CH_3COO)_2.H_2O$, $Ba(CH_3COO)_2.H_2O$, $Sr(CH_3COO)_2.H_2O$ and $Cu(CH_3COO)_2.H_2O$. The homogeneous mixture was calcinated in box furnace at 880°C for 12 hours in order to eliminate the organic residue in sol-gel powder and provide initial reaction to the oxide compound. The powders were palletized with standard diameter of 13 mm and 2 mm thick by hydraulic press at a pressure of 6-8 tons SPECAC. Annealed process were carried out in 2408-Carbilite tube furnace on these pellets form superconductor oxide at temperatures of at 880 °C respectively for 18 hours to make a final bond forming reaction.

3.3. Synthesized of $RuSr_2GdCu_2O_{8-\delta}$ (Ru-1212) Superconductor by Acetate-Tartrate Sol-Gel Route

Starting chemicals used were acetate and sodium chloride salt are completely soluble in weak acid. Metal salts $RuCl_3.H_2O$, $Sr(CH_3COO)_2.H_2O$, $Gd(CH_3COO)_3.H_2O$ and $Cu(CH_3COO)_2.H_2O$ were dissolved in small amount of 0.02 molarity of acetate acid and stirred with a bar magnetic bar stirrer for one hour at temperature below 80 °C. The mixture reaction solution was black in colors due to the present of Ru^{3+} ion. Tartaric acid ($C_4H_6O_6$) with 0.5 ratio to the total number moles of the reaction mixture solution was weight and added into the solution. Tartaric acid acts as a chelating reagent to complex the copper ion before the gelation, otherwise $Cu(CH_3COO)_2 \cdot H_2O$ would readily precipitate during the concentration process in acidic solution. Aqueous solution of ammonium hydroxide (NH_4OH) is added to the black solution to adjust the pH of the solution to 6.0. pH control is essential to prevent crystallization occurred during gelation. The black solution was stirred for several hours at the same temperature, thus hydroxylation and polycondensation of the sol were achieved. The solution was kept at 100 °C until it turned into a black gel. Calcinations were carried out on the sol–gel powders at 950 °C for 24 hours. Then, the black powders were ground and pressed into pellets with standard size by using SPECAC hydraulic press at a pressure of 6-8 tons. The pellets were annealing at 950 °C–1030 °C for 24 hours, followed by cooling step of 1 °C/min until 450 °C and then oven cooled to room temperature. The whole heat treatment process was performed at ambient pressure in oxygen flow as shown in figure 3.7.

The quality of the samples was studied by X-ray diffraction (XRD) analyses with CuK_α radiation ($\lambda=1.5418$ Å) between $2\theta=2°$ and 60° using a Bruker D8 Advance diffractometer. The lattice parameters were obtained from the XRD data using the least squares method. The dc electrical resistance versus temperature measurements (R-T) were carried out using the four-point probe method with silver paste contacts. A CTI Cryogenics Closed Cycle Refrigerator Model 22 and a Lake Shore Temperature Controller Model 330 were used for temperature dependent measurements and the data acquisition system is fully computer controlled. The scanning electron micrographs were obtained using a Philips PV 99 analyzer.

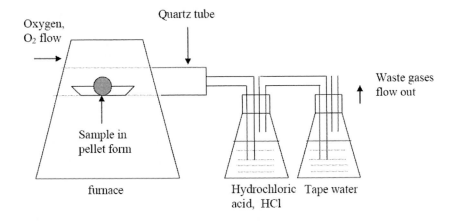

Figure 3.7. Apparatuses arrangement for Ru-1212 annealing process.

3.4. SYNTHESIZED OF RUSR$_{1.5}$CA$_{0.5}$PBCU$_2$O$_{8-\delta}$ (RUPB-1212) SUPERCONDUCTOR BY ACETATE-TARTRATE SOL-GEL ROUTE

The RuSr$_{1.5}$Ca$_{0.5}$PbCu$_2$O$_{8-\delta}$ (Ru-1212) samples were synthesized through the acetate/tatrate sol–gel route with stoichiometric amounts of Ru(Cl)$_3$·H$_2$O ,Sr(CH$_3$COO)$_2$·H$_2$O, Ca(CH$_3$COO)$_2$·H$_2$O, Pb(CH$_3$COO)$_2$·H$_2$O and Cu(CH$_3$COO)$_2$·H$_2$O with 99 % purity. All the chemicals were dissolved in a small amount of distilled water and stirred for several hours at 70 °C. An aqueous solution of tartaric acid (C4H6O6) with a molar ratio to copper fixed at 0.45 was added and the solution was stirred for 1 hour. Tartaric acid acts as a chelating reagent to complex the copper ion before the gelation, otherwise Cu(CH$_3$COO)$_2$H$_2$O would readily to precipitate during the process in acidic acetate solution. To prevent the problem of crystallization of copper acetate during gelation, a dilute ammonium hydroxide, NH$_4$OH solution was added to adjust the pH to 5.5. The solution was stirred for several hours at the same temperature, thus hydroxylation and polycondensation of the sol was achieved. The solution was kept at 100 °C until it turned into a black gel. Calcinations were carried out on the sol–gel powder at 800 °C for 18 hours. Then the black powders were ground and pressed into pellet. The pellet was heated at 850°C–920°C for 24 hours, followed by a cooling step of 1°C/min until 350°C and then furnace-cooled to room temperature. The entire heat-treatment process was performed at ambient pressure in oxygen. The purity of the samples was examined by powder X-ray diffraction (XRD) method with Cu-Ka radiation (λ =1.5418 Å) between 2θ =2°-60° Using a Bruker D8 Advance diffraction meter. Lattice parameters were obtained from the XRD data using the least squares method. The DC electrical resistance versus temperature measurements was carried out using the standard four-probe method with silver paste contacts in conjunction with a CTI closed cycle refrigerator down to13 K. Scanning electron micrographs were obtained using a Philips PV 99 analyzer. The experiment results were show in the chapter IV.

3.5. PREPARATION OF POWDER $RuSr_2(Gd_{2-x}Ce_x)Cu_2O_{8-\delta}$ (RU-1222) THE TECHNIQUE ASG

Starting chemicals used were $Ru(Cl)_3.H_2O$, $Sr(CH_3COO)_2.H_2O$, $Gd(CH_3COO)_3.H_2O$, $Ce(CH_3COO)_2.H_2O$ and $Cu(CH_3COO)_2.H_2O$ that dissolve in weak acetate acid at 80°C. The method of Ru-Sr-Gd-Ce-Cu-O oxide powder preparation by ASG technique was similar with the previous case which is show as Section 3.3 and Section 3.8. Pellets were annealing at temperature of 1050 °C for 24 hours and cooled under the flow of O2 and the controlled cooling at a rate of 1 °C/min up to 200 °C. The sample is allowed to cool naturally in the furnace to room temperature.

3.6. INFRARED SPECTRUM (FTIR) FOR SOL-GEL POWDER

Metal oxide sol-gel powders prepared by acetate-tartrate gel and citric gel routes has been characterized and measure with infrared spectrometer instrument (FTIR) Perkin Elmer model 1605 brand in the School of Chemical Sciences and Food Technology, Universiti Kebangsaan Malaysia.

Around 2mg of sol-gel powder pounded and mixed with 100-200 mg of anhydrous KBr ball precious stones. The mixture were ground and pressed under 10000-15000 psi form a transparent disk. The sample is placed in the FTIR sample. Figure 3.8 shows the spectrum infrared instrumentation.

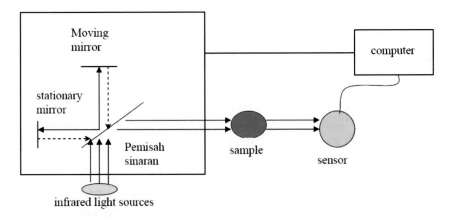

Figure 3.8. Infrared spectrum instrumentation.

Infrared light sources with frequency 5000-400 cm^{-1} split into two grooves. One slot will be made on a stationary mirror and all the radiation will be reflected back to the beam separator. The radiation beam will be reflected on the moving mirror and produce a different wavelength than the first channel. The merger of the two slots in the separator produces radiation pattern, which is a combination of all the interference patterns arising from each of the wavelength of the beam. The merger will be passed to radiation in the sample. Difference in wavelength is absorbed by the sample can be described the presence of certain functional groups. Interference pattern is generated and used computer calculations to convert the interference to a plot of absorption versus wavelengths of infrared spectra similar to normal.

3.7. THERMAL ANALYSIS GRAVIMETRIC (TGA) AND DIFFERENCE THERMAL ANALYSIS (DTA)

In the context of sol-gel chemistry, thermal analysis is a versatile aid to throw light on the nature and the decomposition mechanism of the raw powder [Mathur et al 2002; Tautkus et al. 2000]. Thermal analysis instrument branded Mettler Toledo TGA / SDTA 851 in the School of Chemical Sciences and Food Technology, Universiti Kebangsaan Malaysia were used in this study. This instrument can perform both thermal gravimetric analysis (TGA) and difference thermal analysis (DTA) for same sample at the same time. TGA is use to measure a change in sample weight lost with increasing temperature of treatment. Metal oxide powder prepared by five wet chemistry techniques contain organic materials and water were studied by TGA instrument. 10-15 mg powder samples were weighed and placed into the instrument sample holder. Heating rate was 10 °C/minute from room temperature 30 °C to 800 °C. Graph of sample weight versus temperature or time was recorded by the computer and the percentage of sample weight was obtained. Size of the sample weight measured is within the range of 10-15 mg. Basic components in the thermal analysis instrument is functioning as a recorder balancing thermal balance, furnace, sample holder, temperature programmer, controller and computer with the atmospheric data analysis program. Enthalpy change in a material can be give us the information that is the chemical reaction in the sol-gel powder verse temperature is either exothermic or endothermic process can be measured by DTA.

3.8. MEASUREMENT OF ELECTRICAL RESISTANCE (R) AND CRITICAL TEMPERATURE (T_c)

Electrical resistance and critical temperature of superconducting samples measured by four-point probe method (Figure 3.9).

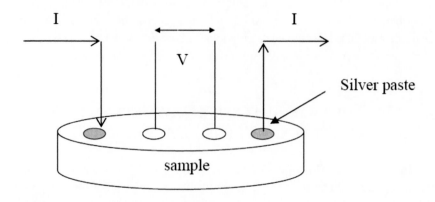

Figure 3.9. Four-point probe method for measuring the electrical resistance and critical temperature.

The sample is attached to the sample holder with Apiezon grease adhesive. Silver paste used for electrical contact between the wire probe with the pellet sample. Silver paste has the lower resistively 0.001 Ωm. During the measurement of the critical temperature, the sample is cooled in vacuum environment in a closed cycle cooling system equipment Model 22 CTI-CRYOGE school board with DT-470 silicon diode as temperature sensor. Sample temperature is controlled by temperature control of the 330th Auto tuning Lakeshore temperature controller. Data on changes in resistance with temperature recorded by the computer. First critical temperature ($T_{c\text{-start}}$) is defined as the temperature at which the electrical resistance began to sort of line that is the temperature at which superconductivity first occurs. Zero critical temperature ($T_{c\text{-zero}}$) is the temperature where the electrical resistance drops to zero as shown in Figure 3.10.

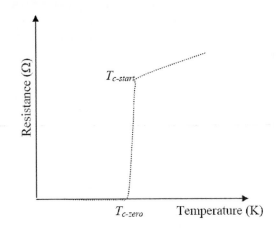

Figure 3.10. Electrical resistance against temperature of a superconducting material.

3.9. X-RAY DIFFRACTION (XRD)

X-ray Diffraction (XRD) is a powerful nondestructive technique for characterizing crystalline materials. It provides information on structures, phases, preferred crystal orientations (texture), and other structural parameters, such as average grain size, crystallinity, strain, and crystal defects. X-ray diffraction on metal oxide powder were use to study the structure and phase of superconductor samples. X-ray diffraction peaks are produced by constructive interference of a monochromatic beam of x-rays scattered at specific angles from each set of lattice planes in a sample. The peak intensities are determined by the atomic decoration within the lattice planes. Diffraction instrument branded Simens D-5000 with Cu Kα radiation and the wavelength λ ≈ 1.5418 Å in the Geology Program, University Kebangsaan Malaysia used. Diffraction angle 2θ from 2° to 60° with spaced 0.04°. Samples in the form of pellets ground into fine powder first before X-ray diffraction. Bragg's law is used to calculate the distance between an atom are:

$$n\lambda = 2d \sin\theta \qquad\qquad (3.1)$$

with n is the order number, λ is the wavelength, θ is the angle of diffraction, and d is the distance between the planes. When a monochromatic X-ray beam with wavelength λ through the sample, the light would diffracted at an angle θ from the original incident ray. A particular diffraction pattern obtained on the crystal structure complies with Bragg on microfilms. Diffraction patterns obtained exhibit intensity peaks at certain angles and angle information θ and the distance between the planes. Intensities peaks are the result of a combination of beam-beam with the same *hkl Miller* index at that point. Figure 3.11 shows the working principle of X-ray diffraction.

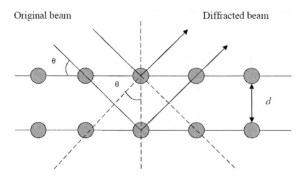

Figure 3.11. X-ray diffraction working principle.

3.10. LATTICE PARAMETER DETERMINATION

In the XRD patterns of samples are recorded, the intensity of the peaks obtained matched with the appropriate Miller indices to identify the structural and superconducting phase of the study. Constants, lattice constant a, b, and c for the sample can be determined by the equation through the program MATHCAD PLUS 6.0.

$$1/d^2 = h^2/a^2 + k^2/b^2 + l^2/c^2 \qquad (3.2)$$

with d the distance between the atoms and h, k, l are Miller indices

REFERENCES

Chen, Y., Yan, F., Zhao, G., Qu, G., Lei, L. 2010. Fluorine-free sol-gel preparation of YBa2Cu3O7-x superconducting films by a direct annealing process. Journal of Alloy and Compounds. 505: 640-644.

Kalubarme, R.S., Shirage, P.M., Iyo, A., Kadam, M.B., Sinha, B.B., Pawar, S.H. 2009. Synthesis and magnetic properties of $Bi_2Sr_2CaCu_2O_y$ Superconductor by Using Nitrate precursors. *J. Superconductor Nov. Magn* 22: 827-831.

Makan, C., Lise, D., Martin, L., Paul, W. 2004. High temperature superconductors for powder applications. *Journal of the European Ceramic Society*. 24: 1815-1822.

Mathur, S., Shen,H., Lecerf, N., Jilavi, M.H., Cauniene, V., Jorgensen., J-E., Kareiva, A. 2002. Sol-gel syhtesis route for the preparation of $Y(Ba_{1-x}Sr_x)_2Cu_4O_8$ superconducting oxide. *Journal of Sol-Gel Science and Technology*. 24: 57-68.

Sarun, P.M., Aloysius, R.P., Syamaprasad, U. 2006. Preparation on high performance (Bi, Pb)-2223 superconductor using a sol-gel synthesized amorphous precursor through controlled gelation. Material letter. 60:3797-3802.

Suryanarayanan, R., Nagarajan, R., Selig, H. & Ben-Dor, L. 2001. Preparation by sol-gel, structure and superconductivity of pure and fluorinated $LaBa_2Cu_3O_{7-d}$. *Physica C* 361: 40-44.

Tautkus, S., Kazlauskas, R. & Kareiva, A. 2000. Thermogravimetric analysis- a powerful tool for the refinement of the synthesis process of Hg-base superconductor. Talanta 52: 189-199.

Zalga, A., Reklaitis, J., Norkus, E., Beganskiene, A., Kareiva, A. 2006. A comparative study of $YBa_2Cu_4O_8$ (Y-124) superconductors preparation by sol-gel method. *Chemical physics* 327:220-228.

Chapter 4

RESULTS, DATA ANALYSIS AND DISCUSSION ON YBCO SUPERCONDUCTOR PREPARATION THROUGH FIVE WET CHEMICAL TECHNIQUES

INTRODUCTION

This section discusses about results and analysis of data obtained from experiments carried out on YBCO system prepared through five different wet chemical techniques. Qualitative results on YBCO sol-gel powder analysis obtained directly through the infrared spectrum (FTIR), thermal gravimetric analysis (TGA) and difference thermal analysis (DTA) [Agimantas et al. 2002]. Standard four-probe technique was used to measure the temperature dependence of resistivity in the range 20-300 K of superconductor sample. X-ray diffraction (XRD) and scanning electron micrographs (SEM) and Energy Dispersive spectroscopy (EDAX) analyzed result was carried on final of superconductor compounds to get the complete results data and use for further characterization. This section included discussions above experimental and analysis results carried out in accordance with their respective series of superconductor system preparation.

4.2. RESULTS AND ANALYSIS OF FIVE WET CHEMICAL TECHNIQUES IN YBCO SUPERCONDUCTOR POWDER PREPARATION

4.2.1. Infrared Spectrum Analysis for YBCO Sol-Gel Powder

Infrared spectrum (IR) results for sol-gel powders prepared by tartarate-acetate sol-gel technique (ASG) and citric gel technique (CT) has been obtained. It is well know that IR analysis of synthesized samples is important for both control of the reaction process and the properties of material obtained. Infrared spectrum gave information about presence of organic ligands and functional groups in the Y-Ba-Cu-O gel powders. Figure 4.1 shows the IR spectrum of Y-Ba-Cu-O sol-gel powders prepared through ASG and CT techniques respectively. By comparison, both IR spectrums are quite similar in terms of peaks absorption and may be divided into four main regions. According to the origin curve graph for precursor gel prepared by CT route, the spectrum show the appearances peaks in certain areas range 850-600 cm^{-1} (814.28, 726.56, 624.68 cm^{-1}) regions which are due to the existence of M-O vibration bond. The absorption at 3100-2800 cm^{-1} are due to the stretching vibrations of CH_3 and CH_2 groups. The presence of CO groups are shown by sharp absorption peak 1406.75 cm^{-1} and the fingerprint at 683.20 cm^{-1}. Broad absorption peak region at 1606.64 cm^{-1} is due to the coordination group, CO-OH. The strong bands observed at 3400-3200 cm^{-1}, 1100-1000 cm^{-1} and the medium intensity bands at 1300-1200 cm^{-1} are probably due to the CH-OH stretching frequencies. A broad absorption in the range 3600-3400 cm^{-1} indicates the presence of –OH functional groups. Water absorption is shown in the range 3100-3700 cm^{-1}. The presence of nitrate group can be seen on top of a small band 1757 cm^{-1} and the absorption peak at 1384.75 cm^{-1}. For ASG curve graph, the peaks of 614.98 cm^{-1} and 672.71 cm^{-1} are indicates the existence of M-O. The main absorption peak 1561.13 cm^{-1} and the fingerprint 680.66 cm^{-1} indicate the presence of CO groups. The present of coordination group, CO-OH is shown at the peak absorption of 1552.52 cm^{-1}. Wide range of absorption at 3408.37 cm^{-1} and the fingerprint on the peak 1045.41 cm^{-1} and 1078.16 cm^{-1}, indicates the presence of –OH functional group. Water absorption in the curve is equal to the ASG curve as shown in the CT range of 3100-3700. The presence of a peak in the middle wavelength 2335 cm^{-1} indicates the presence of functional groups of alkynes -C≡C-. Alkynes group is only found in the ASG curve and does not exist in CT curves. Since both -CH_3

and -CH-OH vibrations were identified, it can assume that both acetate and tartrate ligands are present in the ASG gel powders.

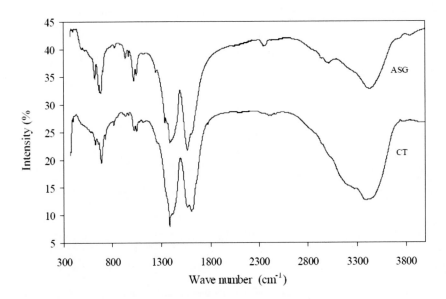

Figure 4.1. Infrared spectrum for Y-Ba-Cu-O sol-gel powder prepared by ASG and CT techniques.

4.2.2. Thermal Analysis Gravimetric (TGA) and Difference Thermal Analysis (DTA) for YBCO Sol-gel Powder

This study analyzes the thermal stability for Y-Ba-Cu-O powder prepared by five different wet chemical techniques. Phase transition temperature for Y-Ba-Cu-O powders can be seen through this analysis. Figure 4.2 (a) shows the TGA/DTA curve graph of Y-Ba-Cu-O gel powders prepared by CT gel technique in the temperature range 30-800°C with a heating rate of 10 °C min^1. From TGA curves, there are three major steps in the sample weight reduction when the temperature increase. The initial weight loss is 33.99 % due to the dehydration was observed in the temperature range 180-230°C, follow by 5.50% weight lost at 230-385 °C temperature range, and the remaining quantity is 49.30 % of 15 mg original sample at temperature of 590 °C. From DTA curves, first endothermic occurred at a temperature of 183.96 °C cause by decomposition of water content and organic materials. A strong exothermic peak occurs at 373.36 °C due to the occurrence of an explosion during the oxidation reaction of the samples. The presence of citric acid and nitric acid in sol-gel powder will cause explosion thus heat release to surrounding.

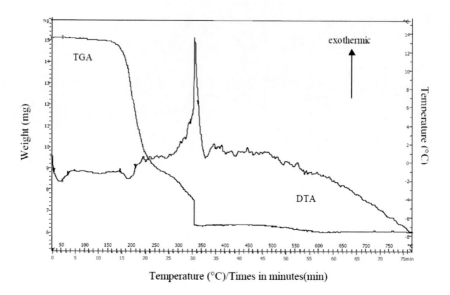

Figure 4.2. TGA/DTA curve graph of Y-Ba-Cu-O powder prepared by CT gel technique.

TGA/DTA curves graph of Y-Ba-Cu-O sol-gel powders prepared via ASG technique in the temperature range 30-800 °C with a heating rate of $10°Cmin^{-1}$ were obtain and shown in Figure 4.3. TGA curve shows four steps in the process of reducing the weight percentage of samples in the temperature range, namely 100-190 °C (1.63%), 190-220 °C (2.68%), 220-280 °C (30.60%), 280-425 °C (19.45%). The weight loss below 175 °C is due to the evaporation of water and organic solvent molecules. From DTA curve, the two significant decomposition steps observed as exothermic feature can attributed to the pyrolysis of organic compounds and the degradation of intermediate species forms during gelation process. An obvious endothermic peak around 200 °C indicates the first decomposition step assignable to removal of absorbed and chemisorbed water. The exothermic feature around 250°C, 300°C and 420°C in the DTA curve are possible due to decomposition of organic legends. Mass of sample powder were established at a temperature 650 °C and the remaining of sample weight are 19.45 % from 15 mg of original sample. This may correspond to begin of YBCO crystallization process and accompanying structure transformation at 650 °C onward.

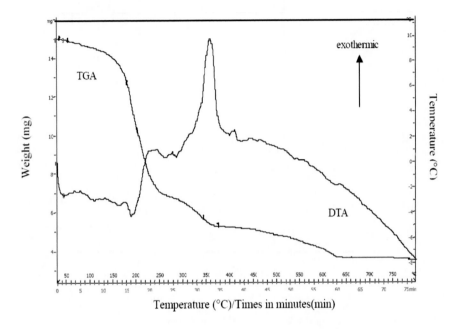

Figure 4.3. TGA/DTA curve graph of powder Y-Ba-Cu-O prepared by ASG technique.

Figure 4.4 shows the TGA/DTA curve graphs of Y-Ba-Cu-O powder prepared by COP technique in the temperature range 30-800°C with heating rate of 10°Cmin^{-1}. From TGA curves, there are only two major steps in the sample weight loss by percentage, namely temperature range 65-270 °C (11.88%) and 270-400 °C (25.80%). Mass of precursor was established at a temperature 700 °C and the remaining of sample weight are 51.29 % from 15 mg of original sample. The result shown that higher calcinations temperature and longer calcinations period are required to obtain the stable value of the weight sample and it may due to thermolysis of organic molecular on the precursor. From DTA curve, the endothermic peak occurs at temperature 145 °C followed by a strong exothermic peak at 364.5 °C.

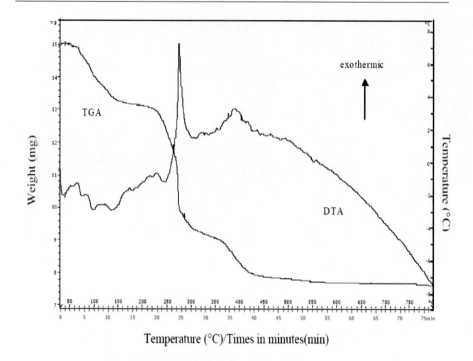

Figure 4.4. TGA / DTA curve graph of Y-Ba-Cu-O powder prepared by COP technique.

TGA/DTA curves graph of Y-Ba-Cu-O powder prepared by ASG-SSR technique at the temperature range 30-800 °C with a heating rate of 10 °Cmin^{-1} shown in Figure 4.5. TGA curve shows three steps in the process of weight loss of sample in certain temperature range, namely 175-235 °C (33.99%), 380-430 °C (5.50%) and 430-515 °C (6.42%). At temperature of 580 °C, the weight of a stable metal oxide powder are obtain with 49.26% of 15 mg original weight. From DTA curves, the first available endothermic occur at a temperature of 216.96 °C due to decomposition of organic material and water content. A broad exothermic reaction in the temperature range 260-640 °C, which takes 38 minutes is most likely caused by bond formation reaction between Ba-Cu-O and Y$_2$O$_3$. In addition, a strong exothermic peak occurred at 468.03 °C and it may due to the oxidation reaction on the samples.

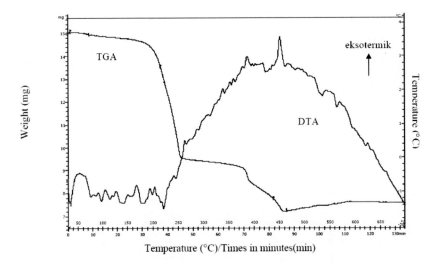

Figure 4.5. TGA / DTA curve graphs of Y-Ba-Cu-O powder prepared by ASG-SSR technique.

4.2.3. Electrical Resistivity against Temperature Changes for YBCO Superconductor Synthesized through Five Wet Chemistry Techniques

Figure 4.6 shows the graphs of change in electrical resistance against temperature for four YBCO samples prepared by sol-gel based on citric gel technique (CT) annealing at temperature of 880 °C, 900 °C, 920 °C, 950 °C for four hours. Sample synthesized at temperature 950 °C shows the nature of the semiconductor at normal state and the rest of another three different annealing temperature show metallic properties at normal state. The optimum annealing temperature was proved by sample synthesis at 900 °C with $T_{c\text{-onset}}$ at 97 K and $T_{c\text{-zero}}$ at 90 K. The narrow transition temperature width, ΔT_c must be due to the fact that the material prepared by this sol-gel route exhibit excellent homogeneity. This means that the temperature of 900 °C is the optimum treatment temperature for YBCO sample preparation via CT technique. The lowest zero critical temperature are obtained for the sample synthesized at 950 °C with $T_{c\text{-zero}}$ 62 K. Samples annealing at 880 °C and 920 °C show $T_{c\text{-zero}}$ value of 84 K and 64 K respectively.

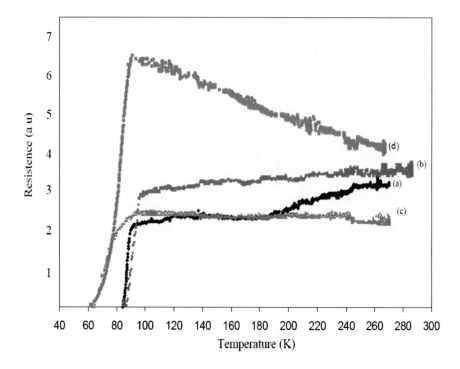

Figure 4.6. Electrical resistance against temperature for the YBCO superconductors synthesized via CT technique with annealing temperature (a) 880 °C, (b) 900 °C, (c) 920 °C, (d) 950 °C.

Figure 4.7 shows the change in electrical resistance against temperature for the YBCO samples pallets prepared by sol-gel technique based acetate tratrate gel route (ASG) annealing at temperature 880 °C, 900 °C, 920 °C, 950 °C for four hours. All samples showed metallic properties at normal state (at room temperature). Sample annealing at 920 °C shows the optimum result with $T_{\text{c-onset}}$ 96 K and $T_{\text{c-zero}}$ of 90 K. This sol-gel route have produce high quality and homogeneity precursor and give the narrow transition temperature width, ΔT_{c}. This means that the temperature of 920 °C is the optimum treatment temperature for ASG technique in the preparation of YBCO samples. The lowest zero critical temperature are obtain for sample annealing at temperature 950 °C with $T_{\text{c-zero}}$ 72 K. Zero critical temperature for sample annealing at 880 °C and 900 °C are 87 K and 84 K respectively.

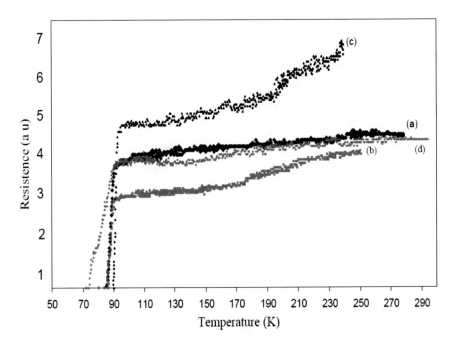

Figure 4.7. Electrical resistance against temperature for YBCO superconductors synthesized via ASG technique with annealing temperature (a) 880 °C, (b) 900 °C, (c) 920 °C, (d) 950 °C.

Figure 4.8 shows the change in electrical resistance against temperature for the YBCO sample prepared by coprecipitation technique, COP. Samples annealing at temperature 920 °C and 950 °C shows metallic properties at while samples annealing at temperature 880 °C and 900 °C show semiconductor properties at normal state (high temperature). Sample annealing at 950 °C shows optimum result with $T_{c\text{-onset}}$ 78K and $T_{c\text{-zero}}$ 87 K. Follow by sample annealing at 920 °C with $T_{c\text{-zero}}$ 84 K Samples annealing at temperature 880 °C only shows onset critical temperature with $T_{c\text{-onset}}$ 62 K and does not achieve zero resistance. The lowest value of zero critical temperature are gain by sample annealing at 900 °C with $T_{c\text{-zero}}$ 35 K. Precursor powder prepared by COP technique are very sensitive to temperature treatment. Annealing temperature exceeding 900°C is required to obtain a quality of superconducting YBCO sample.

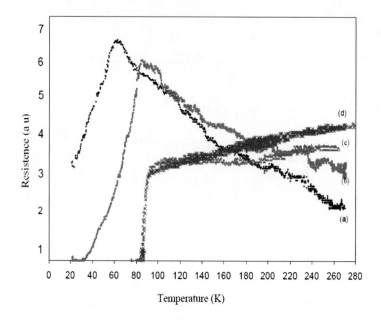

Figure 4.8. Electrical resistance against temperature graphs for the YBCO superconductors synthesized via CT technique with annealing temperature (a) 880 °C, (b) 900 °C, (c) 920 °C, (d) 950 °C.

Figure 4.9 shows the change in electrical resistance against temperature for the YBCO samples prepared by COP-SSR techniques. All samples showed metallic properties at normal state. The highest value of zero critical temperature is achieved by sample annealing at 950 °C with $T_{c\text{-zero}}$ 85 K. This means that 950 °C is the optimum temperature treatment for COP-SSR technique. Sample annealing at 880 °C shows lowest zero critical temperature, with $T_{c\text{-zero}}$ 71 K. samples annealing at 920 °C and 900 °C show $T_{c\text{-zero}}$ 77 K and 84 K respectively. The experiment result shown that this technique have improve the temperature treatment sensitivity compare to COP technique.

Figure 4.10 shows the change in electrical resistance against temperature for the YBCO samples prepared by ASG-SSR technique. Four of the samples shows semiconductor properties at normal state. Annealing temperature at 950 °C is the optimum temperature treatment for ASR-SSR with $T_{c\text{-zero}}$ 83 K. The lowest zero critical temperature are obtain by sample annealing at temperature 880 °C with $T_{c\text{-zero}}$ 70 K. Samples with temperature treatment 920 °C and 900 °C show $T_{c\text{-zero}}$ 74 K and 73K respectively. Table 4.1 shows summary of $T_{c\text{-zero}}$ value for YBCO obtained by the five wet chemistry techniques with particular temperature treatment.

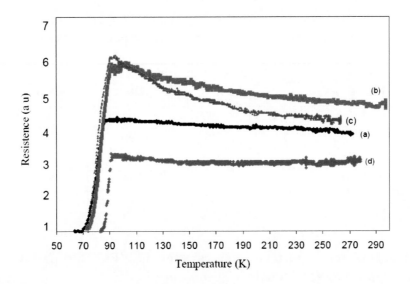

Figure 4.9. Electrical resistance against temperature graphs for the YBCO superconductors synthesized via COP-SSR technique with annealing temperature (a) 880 °C, (b) 900 °C, (c) 920 °C, (d) 950 °C.

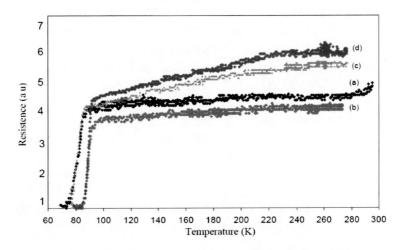

Figure 4.10. Electrical resistance against temperature graphs for the YBCO superconductors synthesized via ASG-SSR technique with annealing temperature (a) 880 °C, (b) 900 °C, (c) 920 °C, (d)950 °C.

Table 4.1. $T_{\text{c- zero}}$ for YBCO obtained by the five wet chemical techniques with different temperature treatment

Techniques	Treatment temperature			
	880 °C	**900 °C**	**920 °C**	**950 °C**
ASG	84	90	64	62
CT	84	87	90	72
COP	--	35	85	87
ASG-SSR	70	74	73	83
COP-SSR	71	77	84	85

4.2.4. Diffraction patterns of X-rays for the YBCO Prepared through Five Wet Chemical Techniques

Figure 4.11 shows X-ray diffraction pattern (XRD) for YBCO samples synthesized by the five types of wet chemical route with optimum annealing temperature. Samples prepared via ASG and CT techniques have shown a single phase 123 in the XRD patterns respectively. The presence of mixed phases 123 and 211 and other phases that are not identified are found in the sample synthesized by COP, COP-SSR and ASG-SSR techniques. Table 4.2 shown the summary of 123 phase percentage and impurity exist in YBCO samples prepared by difference techniques.

Percentage of impurity phase (%) = $\dfrac{\text{number of impurity peaks}}{\text{total number of peaks in XRD pattern}}$ X 100%

Percentage of 123 phase (%) = 100% - percentage of impurity phase (%)

Table 4.2. Summary of percentage of 123 phase and impurity in YBCO samples prepared by difference techniques

Techniques	Percentage of 123 phase	Percentage of impurity
CT	91.67 %	8:33 %
ASG	92.29 %	7.71 %
COP	83.40 %	16.60 %
COP-SSR	80.00 %	20.00 %
ASG-SSR	75.00 %	25.00 %

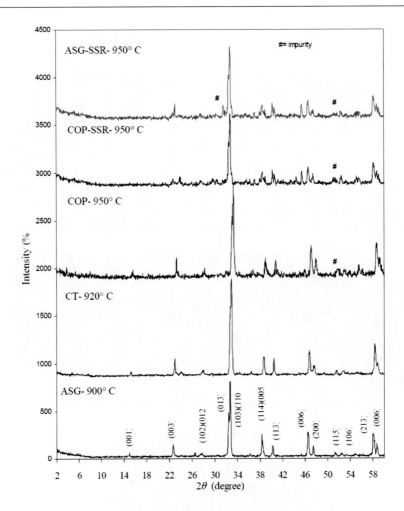

Figure 4.11. Comparison of XRD pattern for YBCO samples prepared by ASG ,CT, COP, ASG-SSR and COP-SSR techniques.

4.2.5. Scanning Electron Micrographs (SEM) for YBCO Samples prepared via Five Wet Chemical Techniques

Figure 4.12 indicates the SEM micrographs of the resulting YBCO samples prepared via these five wet chemical route. SEM tests with the power expansion of 1:00 KX and 5:00 KX been carried out on YBCO samples prepared by five wet chemical techniques. Figure 4.12 (a), (b), (c), (d) and (e) indicates the microstructure of YBCO sample synthesis by ASG, CT, COP, COP-SSR and ASG-SSR respectively at the optimum annealing temperature.

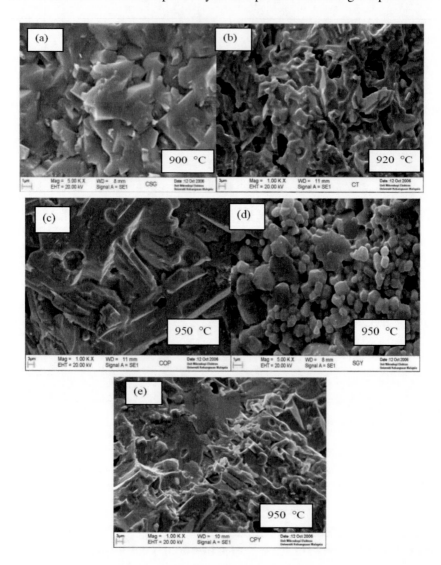

Figure 4.12. SEM micrographs for the YBCO samples synthesis via (a) ASG, (b) CT, (c)COP (d) ASG-SSR and (e) COP-SSR.

4.3. DISCUSSION

The discovery of high-temperature cuprate-based superconductor Y–Ba–Cu–O in 1987 has stimulated the intensity of scientists in development of the superconducting application theoretically. Nowadays the search for high temperature superconductivity and novel superconducting mechanisms is one of the most challenging tasks of material scientists. As a consequence, numerous approaches and efforts have been tried to synthesize oxide superconducting powders as pure as possible with optimal morphology and physical properties. Solid-state technique is the most popular route to prepare $YBa_2Cu_3O_{7-\delta}$ [Kirschener et al. 1997]. It involves the use of high purity oxides and carbonate powders as starting materials. Multiple grindings and extended heat treatment are necessary to achieve the complete reaction, since the formation of the superconducting phase proceeds via diffusion in the solid state. However, higher annealing temperature and longer heating duration would bring about the evaporation of the powders, and consequently lower the quality of the sample [Mishra et al. 1992]. To improve the quality of the samples several chemical preparation techniques, such as the sol–gel [Peng et al. 2004] and coprecipitation technique [Hamadneh et al.1995], have been developed to achieve better mixing of the initial products. In this work, four types of wet chemical preparations of superconducting oxides were investigated. When metals are mixed in a solution form in the acetate–tartrate sol–gel route (ASG) and the complexion citric gel route (CT) an excellent homogeneous oxide precursor for YBCO can be produced. Low sintering temperature and shorter duration are required in these sol–gel methods for YBCO preparation. The problem of solubility in certain metal salts limits the usage of the wet chemical route. This sol–gel-solid-state technique (ASG-SSR) is designed to overcome the weaknesses in sol–gel method (because of the difficulty in dissolving or indissolubility for these metal oxides in water or acid). The organic compounds, such as –OH group, present in Ba–Cu–O powder are chemically reactive to Y_2O_3 and have the potential to provide extra oxygen to stabilize the superconducting phase. Coprecipitation route (COP) has the advantage of producing ultrafine submicrometer oxide powders. Comparison of superconducting oxides produced by these four wet chemical routes and the final product will be based on the thermal gravimetric and differential thermal analysis (TGA/DTA), FTIR, electrical resistance (R-T), X-ray diffraction pattern (XRD), and scanning electron microscopy (SEM)

4.3.1. Characterization on Sol-Gel Powder with FTIR Analysis and TGA / DTA Thermal Analysis

FTIR results for the Y-Ba-Cu-O gel powder prepared via CT gel route and ASG gel route are quite similar in terms of qualitative. Both curve graphs have similar with four main areas. Range of 3600-3400 cm^{-1} indicates the presence of –OH functional groups. Water absorption is shown in the range of 3100-3700 cm^{-1} while 850-600 cm^{-1} regions are indicates the presence of M-O bond. -COOH group is shown in the 1500-1606 cm^{-1} region. A small band peak at 1757 cm^{-1} which is due to nitrite group is only exist in the CT gel curves graph and peak at 2335 cm^{-1} is indicate the presence of alkyl group which is only present in ASG gel curve graph.

Thermal analyses are carried out on Y-Ba-Cu-O powder prepared by CT, ASG, ASG-SSR and COP techniques via TGA and DTA test. Thermogravimetric analysis (TGA) is a thermal analysis technique which measures the weight change in a material as a function of temperature and time, in a controlled environment. This can be very useful to investigate the thermal stability of a material. As comparison, samples prepare by COP take longer and higher calcinations temperature to get a stable residue weight percentage, it need to heat at 700 °C to remove the entire residue organic compound. Sample prepared by ASG contain high percentage of organic compound, there is 75.81 % and it only achieve stable weight percentage at 650 °C.

Table 4.3. Percentage of stable residue weight of metal oxide in sample when achieve its crystalline temperature

Techniques use to prepare YBCO powder	Percentage of weight residue metal oxide in the simple (%)	Temperature to achieve stable weight percentage of metal oxide (°C)
Citric gel (CT)	49.30	590
Acetate-tratrate gel (ASG)	24.19	650
Coprecipitation (COP)	51.29	700
Sol-gel-solid state reaction (ASG-SSR)	49.26	580

Differential thermal analysis (DTA) is a calorimetric technique, recording the temperature and heat flow associated with thermal transitions in a material. DTA graphs for the four samples shows a peak at endothermic temperature range 140-220 °C. Endothermic peak is due to evaporation of water molecules and decomposition of organic solvent in the sample. Sample prepared by COP technique found to have the lowest hydration temperature at which the temperature is 140.5 °C. Sample prepared by both ASG and CT sol-gel techniques of have quite similar in their thermal analysis and show an endothermic peak at temperature 184 °C and 193 °C respectively. Sample from ASG-SSR technique showed an endothermic peak at the highest temperature compare to the rest, ie 217 °C. This may be due to steric effects arise from the fact that the presences of Y_2O_3 compound occupies a certain amount of in the homogeneous mixture of precursor which will prevent the evaporation of water molecules. Higher heat energy are required absorbed by the system to enable water molecular escape from the sample.

An exothermic peak is clearly seen in DTA curve graphs for these samples prepared by these four difference techniques. This may due to oxidation reactions in the sample. Heat released from the system for new bond formation in the samples. ASG, CT and COP techniques showed a strong exothermic peak respectively in the temperature range 349-373 °C. A broad exothermic band are occurred in ASG-SSR sample and this may due to the processes in this system is relatively slow because the oxidation reaction, diffusion process and the formation of new bonds between -Ba-O-Cu-O-and Y_2O_3 rather complicated. Figure 4.13 shows the comparison on exothermic and endothermic for these YBCO samples prepared by four wet chemical techniques.

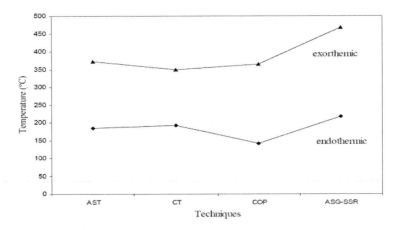

Figure 4.13. Comparison of exothermic and endothermic temperature for YBCO samples prepared by four wet chemical techniques.

4.3.2. Superconductivity

Superconductivity is the analysis of temperature dependence of the resistivity in the YBCO samples synthesized by five different wet chemistry techniques. Figure 4.14 shows the graph changes $T_{c\text{-zero}}$ value dependence on four different annealing temperature treatment for five wet chemical techniques. Samples prepared by both sol-gel techniques, namely CT and ASG required low calcinations temperature to stabilize its YBCO system and get highest $T_{c\text{-zero}}$. Optimum calcinations temperature for four hours on sample prepared by CT and ASG are 920 °C and 900 °C respectively. For ASG techniques, condensation occurs during gelation and the final result is the formation of homogeneous polymerization chain -$[Y\text{-}O\text{-}Ba\text{-}O\text{-}Cu\text{-}O]_n$- sol-gel powder therefore low and short calcinations temperature are enough to establish a stable YBCO crystalline structure [Brinker & Scherer 1990]. In the case of CT gel, metal ions are distributed homogeneously in polymer gel produced through esterification between ethylene glycol and citric acid [Kakihana 1996]. 300 °C of calcination temperature have been able to decompose all the polymer gel and organic solvent, thus further increase the calcinations temperature is use to stabilize the orthorhombic structure of YBCO superconductor. In both sol-gel techniques, low annealing temperature is more suitable to get a quality superconductor material. The high annealing temperature will cause the evaporating of some metal components and destroy the stoichiometry of YBCO structure.

Instead, COP, COP-SSR and ASG-SSR techniques require higher annealing temperature to stabilized the orthorhombic structure of YBCO superconductor and obtain highest $T_{c\text{-zero}}$ value compare to sol-gel techniques. COP technique preparation is base on the concept of separation solid phase containing various species of cations from solution phase. COP technique would produced a heterogeneous solid powder precursor comprising of fine particles typically in the range between 100-500 nm, that greatly reduces the diffusion distances compared with those required for the solid-state reaction route, resulting in shorter reaction times and lower reaction [Pathak and Mishra 2005]. 950 °C is the optimum annealing temperature for COP, COP-SSR and ASG-SSR within four hours durations. Formation of superconductor materials in this three partial solid-state route are based on the process of diffusion and infiltration of heterogeneous powders during heat treatment. $T_{c\text{-zero}}$ value of YBCO obtain from these three techniques are directly proportional to the increase of annealing temperature. The studies shown that $T_{c\text{-zero}}$ value through ASG-SSR and COP-SSR techniques does not have a significant shift compared with the COP technique. This is because the presence of organic material in ASG and COP powders precursor enable to enhance infiltration and diffusion rate during the oxidation process with Y_2O_3. Samples synthesized at temperatures 880 °C with techniques ASG-SSR and COP-SSR show $T_{c\text{-zero}}$ 70 K and 71 K respectively while sample synthesized through COP techniques only shows $T_{c\text{-onset}}$ 50 K and does not indicate superconductivity properties. Results showed that samples prepared through the COP technique is very much sensitive to temperature treatment compare to the rest. Treatment temperature exceeding 900 °C is more suitable for COP route to obtain quality of superconducting material.

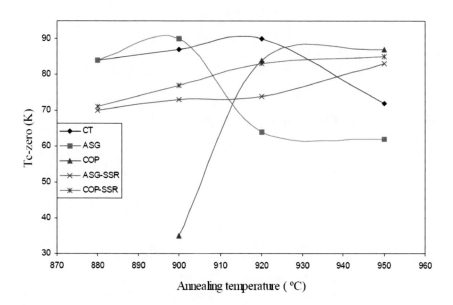

Figure 4.14. Graph changes of critical temperatures, $T_{c\text{-zero}}$ against annealing temperature within four hours for five wet chemical techniques.

4.3.3. YBCO Phase Purity and Lattice Parameters

Referring to Figure 4.11, YBCO samples synthesized through by ASG and CT techniques show a single phase 123. XRD pattern for samples synthesized through COP, COP-SSR and ASG-SSR show the presence of impurity phase, anyway orthorhombic structure are still remain. This is because the annealing temperature and duration of treatment of these samples are not in optimum condition. Table 4.3 shows the value of $T_{\text{c-onset}}$, $T_{\text{c-zero}}$ optimum and lattice parameters a, b and c for samples synthesized via five wet chemical techniques.

Table 4.3. The summarize of $T_{\text{c-onset}}$, $T_{\text{c-zero}}$, lattice parameter for YBCO according to five wet chemistry techniques

Techniques	$T_{\text{c-onset}}$ (K)	$T_{\text{c-zero}}$ (K)	a (Å) ($\pm 0,001$)	b (Å) ($\pm 0,001$)	c (Å) ($\pm 0,001$)
CT	92	90	3.851	3.788	11.433
ASG	90	87	3.853	3.787	11.448
COP	89	85	3.860	3.784	11.466
COP-SSR	88	85	3.859	3.799	11.445
ASG-SSR	86	83	3.861	3.800	11.423

4.3.4. Scanning Electron Micrographs (SEM)

Scanning Electron Microscopy (SEM) focused electron beam across a sample surface, providing high-resolution and long-depth-of-field images of the sample surface. Figure 4.12 shows the morphology of YBCO sample synthesized through the five wet chemical techniques are respectively treated with the optimum annealing temperature. SEM for samples prepared by technique CT 4.12 (b) has a better layered type structure with stronger link compare to the rest. This is an indication of much better link between the superconducting grains. Sample prepared by ASG also show good grains size but there show some melting evidence and agglomerated grains have been observed in the microstructure of sample shown in Figure 4.12 (a). Both CT and ASG offer films and fibers to be form directly from the gel state. The grain alignment of sample prepared by COP route has higher density compare to the rest and suitable for tape application. Samples prepared by ASG-SSR is composed of uniform small rounded grains, pores and shows no evidence of melting even it is sintering at 950 °C

REFERENCES

Algimantas, B., Darius, J., Aivaras, K. 2002. Characterization of sol-gel process in the Y-Ba-Cu-O acetate-tartrate system using IR spectroscopy. *Vibrational Spectroscopy* 28: 263-275.

Brinker, C.J. & Scherer. G.W. 1990. *Sol-Gel Science, The physics and chemistry of sol-gel processing.* New York: Academic Press.

Kakihana, M. 1996. Sol-gel preparation of high temperature superconducting oxides. *Journal of Sol-Gel Science and Technology* 6: 7-55.

Kirschner, I., Bodi, A.C., Laiho, R., Lahderanta, L. 1997. *J.Mater.Res.* 12: 3090.

Mishra, S.K., Pathak, L. C., Rao, M.V.H., Bhattacharya, D., Chopra, K.L. 1992. *Indian L. Pure. Appl. Phys.* 30:685.

Peng, C. H., Hwang, C.C., Chen, S.Y. 2004. *J.Mater.Sci.* 39: 4057.

Chapter 5

RESULTS, DATA ANALYSIS AND DISCUSSION ON DOPING AND ADDITIONAL IN YBCO SYSTEM SUPERCONDUCTOR

INTRODUCTION

This chapter discusses results and analysis of data obtained from experiments carried out on superconducting YBCO system. Research studies conducted on this topic are the addition effect of nano size argentums and the effect of Ca and Sr doping on YBCO system respectively. Data results obtained directly through the infrared spectrum (FTIR) to the sol-gel powder prepared by the ASG technique. After that, standard four-probe technique was used to measure the temperature dependence of resistivity in the range 20-300 K of superconductor sample (RT), x-ray diffraction (XRD), Energy Dispersive spectroscopy (EDAX) and scanning electron micrograph (SEM) are done on the samples to get the complete results data and use for further characterization.

5.2. RESULTS AND ANALYSIS ON YBCO + N% NANO ARGENTUM SERIES

5.2.1. Infrared Spectrum of Sol-Gel Powder Ba-Cu-O

To understand the bonding of the metal ions in the gels and the mechanism of the sol-gel process, Infrared spectrum (IR) tests were conducted on the Ba-Cu-O sol-gel powder prepared by ASG technique. From the FTIR results observed in Figure 5.1, there are four main regions that are quite similar as results in Figure 4.13. Peaks at 618.98 cm^{-1}, 655.81 cm^{-1} and 766.53 cm^{-1} are attributed to M-O. The main peak at 1415.63 cm^{-1} and the fingerprint at 680.66 cm^{-1} are indicated the presence of CO in the sample. The exists of main coordination group -COOH is shown at the peak of 1552.52 cm^{-1}. There is a peak with wavelength 2328 cm^{-1} indicate the presence of alkynes functional group -C≡C-. At the main peak at 3413.37 cm^{-1} and the 1094.41 cm^{-1} and 1049.16 cm^{-1} fingerprints indicate the presence of -OH functional groups. In addition, a broad water absorption is shown in the range of 3100-3700 cm^{-1}.

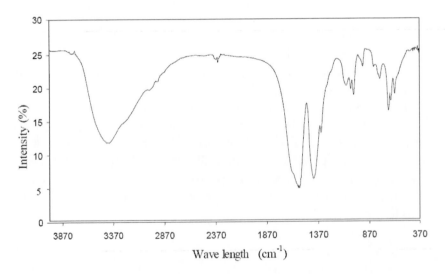

Figure 5.1. Infrared spectrum of the powder sol-gel Ba-Cu-O prepared by ASG technique.

5.2.2. Electrical Resistivity against Temperature Changes for YBCO + n% Nano Ag Series

Figure 5.3 shows the change in electrical resistance against temperature for a series of YBCO + n % nano Ag with $n = 0, 3, 5, 10, 15$ prepared through ASG-SSR. $T_{c\text{-onset}}$ and $T_{c\text{-zero}}$ of these series of five YBCO Ag added samples are obtained respectively in the range between 91 to 94 K and 78 to 90 K. Addition of 15% nano Ag shows the lowest with $T_{c\text{-zero}}$ 78 K compared to $T_{c\text{-zero}}$ 90 K obtained by pure YBCO. $T_{c\text{-zero}}$ value achieved by the addition of 2%, 5% and 10% nano Ag sample are respectively at 83 K, 84 K and 85 K. Figure 5.2 shows the variation of onset temperature, $T_{c\text{-onset}}$ and zero-resistance temperature, $T_{c\text{-zero}}$ against % weight of nano Ag. Additional of nano Ag on YBCO does not substantially influence $T_{c\text{-onset}}$ for this series of five samples. However, the addition of nano Ag will counteract the zero-resistance temperature with the lowest value obtained for 15% weight of nano Ag-addition which showed $T_{c\text{-zero}}$ of 78 K. When the percentage of Ag increased beyond a certain limit, the Y_2BaCuO_5 (Y-211) and $BaCuO_2$ phase increased and suppressed the superconducting behavior of YBCO. The high percentage of Ag which exists between the crystalline structures of YBCO has obstructed the freedom of the 'supercurrent' in the sample. The normal state for resistance-temperature curves graph for all the Ag-added samples exhibited a metallic behavior, except the 15%-Ag weight YBCO sample which is shows semiconducting behavior at normal state.

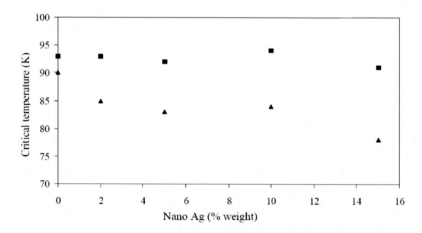

Figure 5.2. The change of $T_{c\text{-onset}}$ and $T_{c\text{-zero}}$ value against YBCO+ n% nano Ag series.

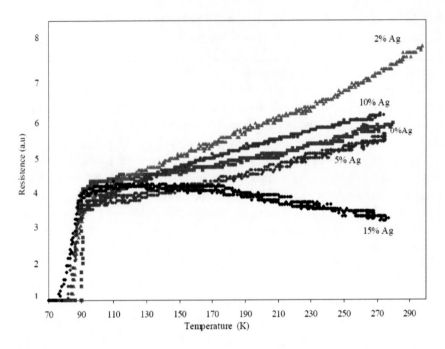

Figure 5.3. Electrical resistance against temperature for a series of YBCO + n % nano Ag with $n = 0, 3, 5, 10, 15$ prepared through ASG-SSR.

5.2.3. X-ray Diffraction Patterns for YBCO + $n\%$ Nano Ag

X-ray diffraction patterns (XRD) for a series of mergers YBCO + n % nano Ag with $n = 0, 2, 5, 10, 15$ with annealing temperature 880 °C for 18 hours has been shown in Figure 5.4. The non-added YBCO sample consists of single phase of Y-12. Other than the main phase 123, XRD pattern for YBCO samples with 2% and 5% weight of nano Ag additional has shown the presence of foreign peaks caused by the presence of Y_2BaCuO_5 (Y-211) and $BaCuO_2$ phase. Impurities are more obvious when the weight percentage of nano Ag added increased to 10 %. Impurities that are Y-211, $BaCuO_2$ and Ag peaks are obviously shown in the sample of 15% weight Ag added. However, the present of impurity phase due to the nano Ag additional does not change the orthorhombic structure in YBCO samples because results based on lattice parameters obtained for the five samples that is $a \neq b \neq c$ as shown in Table 5.1.

**Table 5.1. Lattice parameters and volume
for orthorhombic YBa₂Cu₃O₇ + n% nano Ag**

n value	a (Å) (± 0.001)	b (Å) (± 0.001)	c (Å) (± 0.001)	Volume (Å³) (± 0.001)
0	3.855	3.788	11.433	166.95
2	3.859	3.786	11.451	167.30
5	3.864	3.784	11.469	167.69
10	3.681	3.799	11.412	159.59
15	3.688	3.781	11.410	159.10

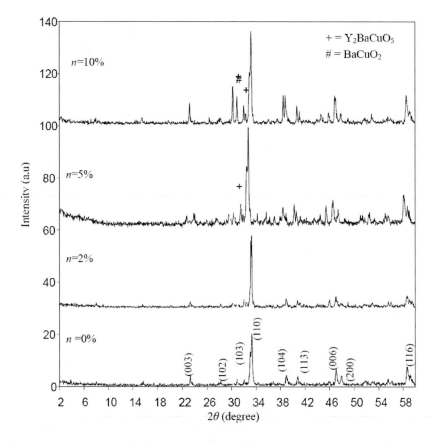

Figure 5.4. The X-ray diffraction patterns (XRD) for the merger series of YBCO + n% nano Ag with n = 0, 3, 5, 10.

5.2.4. Current Density (J_c) for YBCO

Figure 5.5 shows the dependence of critical current density, J_c on nano Ag content at 77 K in YBCO superconductor. J_c value increased dramatically from 1.0 A/cm^2 for pure YBCO to 1.5 A/cm^2 when the 2% nano Ag added. J_c value increase slowly for YBCO-5% Ag nano with 1.7A/cm^2 follow by YBCO-10% obtain the highest J_c of 2.2 A/cm^2. However, when the nano Ag increase to 15%, superconductivity of the sample will decrease with $T_{c\text{-zero}}$ only occur at 78 K, the J_c value also decrease dramatically.

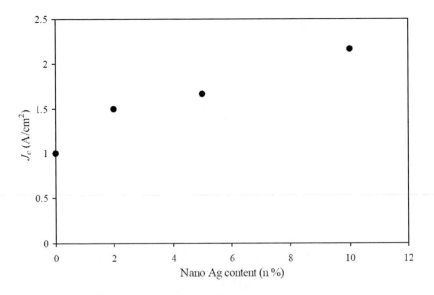

Figure 5.5. The temperature dependence of J_c on Ag content in YBCO at 77 K.

5.2.5. Scanning Electron Micrographs (SEM) and EDAX for YBCO+ n% Ag

Figure 5.6 (a) shows the scanning electron micrographs (SEM) on the nano size of Ag used in this work. Figures 5.6 (b)–5.6 (e) show the microstructure of YBCO+ n % Ag with n=0, 2, 5, 10, 15 respectively. The SEM revealed some changes in the microstructure with different nano Ag concentrations. The pure YBCO composed of small rounded grains, pores and microcracks. When nano Ag are added, the particles are uniformly distributed. EDAX results show that the existence of all metal components in accordance with the expected ratio of yttrium, barium, copper and argentums in the sample with a high oxygen content.

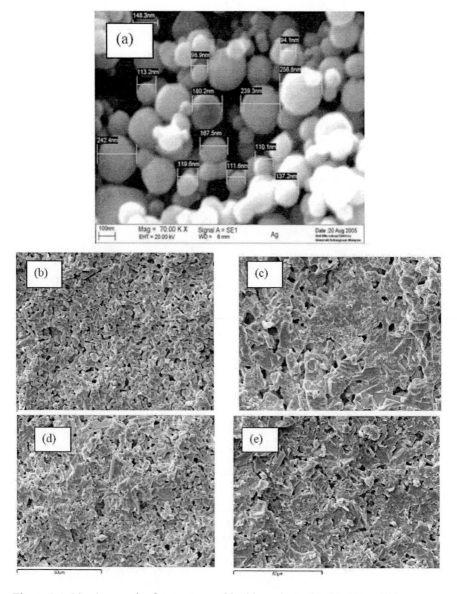

Figure 5.6. (a) micrograph of nano Ag used in this project. (b), (c), (d) and (e) distribution for a series of YBCO +n % Ag with n = 0, 3, 5, 10.

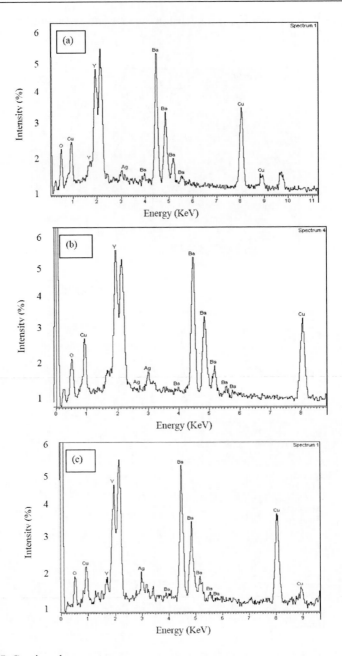

Figure 5.7. Continued on next page

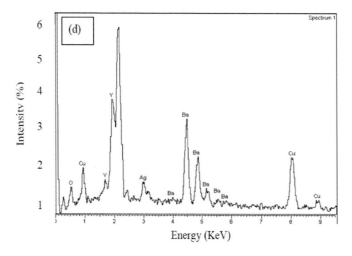

Figure 5.7. EDAX analysis of (a) YBCO, (b) YBCO + 2% nano Ag, (c) YBCO nano +5% Ag, (d) YBCO + 10% nano Ag.

5.3. RESULTS AND ANALYSIS SYSTEM FOR $Y_{0.9}CA_{0.1}BA_{1.8}SR_{0.2}CU_3O_{7-\Delta}$

5.3.1. Infrared Spectrum of the Ca-Ba-Sr-Cu-O Sol-Gel Powder

To understand the bonding of the metal ions in the gels and the mechanism of the sol–gel process, infrared (IR) spectra of the Ca–Ba–Sr–Cu–O precursor gel were measured which is presented in Figure 5.8. A broad and multiple absorptions in the range 3600–3400 cm^{-1} can be assigned to O–H group (intramolecular hydroxyl group), a specific peak have found at 3413.37 cm^{-1}. According charge versus pH diagram [Kepert 1972; Kakihana 1996], low valence cations ($Z < +4$) yields aquo, hydroxo, and aquo or hydroxo complexes over the complete pH scale. The process of gelation in the system investigated probably occurs via the olation mechanism, which involved by nucleophile substitution, SN process where the M–OH or complexing ligands is nucleophile and H_2O is leaving group. In addition, the observed specific IR peaks at 766.53, 680.66, 662,1 and 618.98 cm^{-1} may be attributed to characteristic O–M vibrations [Baranauskas 2001]. From this point of view, it shows that metal-bound intramolecular –[Ca–O–Ba–O–Sr–O–Cu]$_n$– is probably present in the precursor gel, where n is the infinity number. A broad absorption in the spectrum of the precursor gel in the range 3100–3700 cm^{-1}

can be indicated the presence of adsorbed water. The strong bands are due to CO–OH stretching can be identified at range 3500-3200 cm^{-1} and fingerprint region at specific peaks at 1552.54, 1346.4 and 1415.63 cm^{-1}.

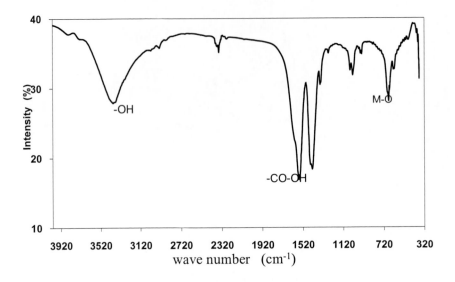

Figure 5.8. Infrared spectrum of Ba-Sr-Ca-Cu-O powder prepared by ASG technique.

5.3.2. Electrical Resistivity against Temperature Changes of the $Y_{0.9}Ca_{0.1}Ba_{1.8}Sr_{0.2}Cu_3O_{7-\delta}$ Superconductor

Figure 5.9 shows the change in electrical resistance against temperature for $Y_{0.9}Ca_{0.1}Ba_{1.8}Sr_{0.2}Cu_3O_{7-\delta}$ superconductor prepared by ASG-SSR techniques. Samples showed metallic behavior at normal state (room temperature) with $T_{c-onset}$ 86 K and T_{c-zero} 80 K. Although the superconducting transition temperature for this bulk superconductor is slightly higher than the boiling point of liquid nitrogen (77 K), it showed decrement as compared to that of pure YBCO with T_{c-zero} value of 90 K. Figure 5.10 shows the comparison graph of electrical resistance against temperature changes on $Y_{0.9}Ca_{0.1}Ba_{1.8}Sr_{0.2}Cu_3O_{7-\delta}$ and YBCO prepared by ASG-SSR technique.

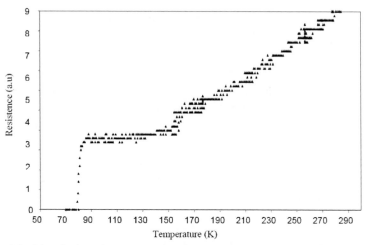

Figure 5.9. Electrical resistance against temperature for $Y_{0.9}Ca_{0.1}Ba_{1.8}Sr_{0.2}Cu_3O_{7-\delta}$ prepared by ASG-SSR techniques.

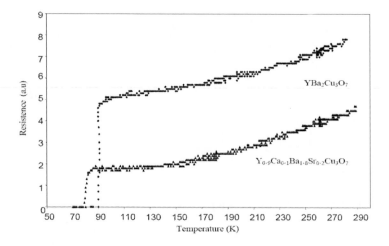

Figure 5.10. Comparison of electrical resistance against temperature for $Y_{0.9}Ca_{0.1}Ba_{1.8}Sr_{0.2}Cu_3O_{7-\delta}$ and YBCO respectively prepared by ASG-SSR technique.

The doping of Calcium (Ca) and Strontium (Sr) which have small radii into Yitrium (Y) and Barium (Ba) respectively will distort lattice parameter and construct the volume structure thus the flow path of supercurrent also disturbed.

5.3.3. Comparative X-rays Diffraction pattern for $Y_{0.9}Ca_{0.1}Ba_{1.8}Sr_{0.2}Cu_3O_{7-\delta}$ and YBCO

X-ray diffraction patterns (XRD) for samples of $Y_{0.9}Ca_{0.1}Ba_{1.8}Sr_{0.2}Cu_3O_{7-\delta}$ and YBCO is shown in Figure 5.11.   Pure YBCO sample consists of 123 single phase.   XRD pattern for sample $Y_{0.9}Ca_{0.1}Ba_{1.8}Sr_{0.2}Cu_3O_{7-\delta}$ shown foreign phase owned by the phase Y_2BaCuO_5 (Y-211) and $BaCuO_2$ peaks other than the main phase 123.   However, the orthorhombic structure of samples retained. X-ray data analysis found that increasing the a and c lattice parameters with doping of Ca and Sr in the Y and Ba for YBCO sample as shown in Table 5.2.

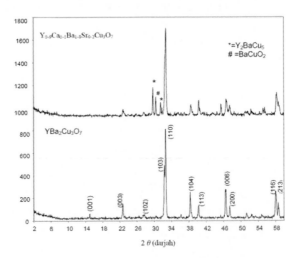

Figure 5.11. Comparison of X-ray diffraction patterns (XRD) for YBCO and $Y_{0.9}Ca_{0.1}Ba_{1.8}Sr_{0.2}Cu_3O_{7-\delta}$.

Table 5.2. Lattice parameter and impurity phase for YBCO and $Y_{0.9}Ca_{0.1}Ba_{1.8}Sr_{0.2}Cu_3O_{7-\delta}$

Sample	a (Å) ±0.001	b (Å) ±0.001	c (Å) ±0.001	%123 ± 0.01	%211 ± 0.01	%BaCuO2 ± 0.01
YBa2Cu3O7-δ (YBCO)	3.824	3.886	11.681	>97.00	0.00	0.00
Y0.9Ca0.1Ba1.8Sr0.2 Cu3O7	3.887	3.829	11.690	66.32	20.00	13.69

5.3.4. Scanning Electron Micrograph (SEM) and EDAX for $Y_{0.9}Ca_{0.1}Ba_{1.8}Sr_{0.2}Cu_3O_{7-\delta}$

The scanning electron micrographs for pure YBCO and bulk $Y_{0.9}Ca_{0.1}Ba_{1.8}Sr_{0.2}Cu_3O_{7-\delta}$ samples annealing at 880C are shown in Figure 5.12 (a) and (b). SEM for YBCO sample showed fine, sharp grains 10 μm in size with no evidence of melting. The size of grain and porosity of sample decrease and these grains were agglomerate for the bulk $Y_{0.9}Ca_{0.1}Ba_{1.8}Sr_{0.2}Cu_3O_{7-\delta}$. Evidence of melting can be observed in the SEM. It can conclude that low temperature is required for this method.

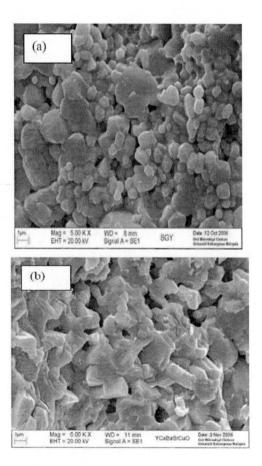

Figure 5.12. Scanning electron micrograph for (a) YBCO and (b) $Y_{0.9}Ca_{0.1}Ba_{1.8}Sr_{0.2}Cu_3O_{7-\delta}$.

Figure 5.13 shows that EDAX and percentage of atomics present in bulk superconductor $Y_{0.9}Ca_{0.1}Ba_{1.8}Sr_{0.2}Cu_3O_{7-\delta}$ compound. Excellence stoichiometry had been obtained for this bulk sample. We found that the average oxygen content of this bulk superconductor was determined to be 6.81 where $\delta = 0.19$. For error estimates, this EDAX result was taken randomly from the surface of sample and there was no guarantee that the whole area of sample is perfect in balance stoichiometry.

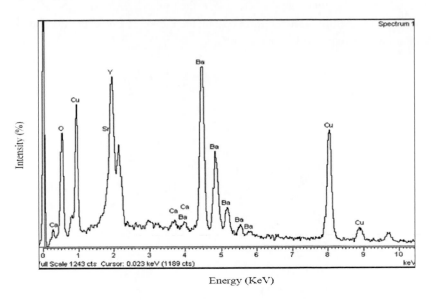

Figure 5.13. EDAX analysis for $Y_{0.9}Ca_{0.1}Ba_{1.8}Sr_{0.2}Cu_3O_{7-\delta}$ showed the existence of all elements with the right ratio.

5.4. DISCUSSION

5.4.1. Effect of the Addition of Nano Ag YBCO

Recently, the role of silver additional of YBCO superconductor had been widely investigates in order to overcome this problem. Many report show that these poor mechanical and electrical properties can be improve by the silver adding [Joo et al.1998; Chuang et al. 1995; Orlava et al. 1998; Fan et al. 1997; Tepe et al. 2004; Costa et al. 2002]. Generally, it is believe that silver will diffuse into the grain boundary as a metal, thus increasing interconnections between the grains and make the conduction easier [Sato et at. 2002]. It will cause the amount of weak links in the structure to decrease and improve the pinning centers thus improving the transport critical current density. The silver can be introduced in various ways such as by mixing with metallic Ag [Joo et al. 1998, Sato et al. 2002, Mathur et al. 2002], mixing with Ag_2O [Joo et al.1999, Karppinen 1995, Zhao et al. 2005], mixing with Ag_2O_2 [Orlava et al.1998], $AgNO_3$ [Joo et al 1998] and by electrochemical methods [Görür et al. 2005]. So far, not many work on the nano size metallic Ag addition on $YBa_2Cu_3O_{7-\delta}$ has been report. We synthesized both monolithic $YBa_2Cu_3O_{7-\delta}$ and composite $YBa_2Cu_3O_{7-\delta}$-Ag through sol-gel-solid-state method

A) The Nature of Superconductivity

Critical temperature of samples changed systematically with the increasing of nano Ag percentage. Figure 5.14 shows the electrical resistance against temperature for samples YBCO + $n\%$ Ag with $n = 0, 2, 5, 10, 15$ nano argentums by weight. Addition of nano Ag in YBCO found not much affect the value of $T_{c-onset}$, however, will cause decreasing of zero critical temperature, T_{c-zero} of the samples. This is because the addition of argentums as impurities in the YBCO samples will cause the production of Y-211 and $BaCuO_2$ phase and the percentage of Y-123 phase has been reduced, thus construct the zero critical temperature. Results on the effect of nano Ag addition is consistent with the effect of addition of Ag metal [Joo et al. 1998], Ag_2O [Abdelhadi et al. 1993] and Ag_2O_2 [Sato et al. 2002 in YBCO. However, the results show that the presence of nano Ag in low quantities have ability to fix the vortex when superconductivity without damage orthorhombic structural in YBCO. Therefore, the path of Cooper pairs are not restricted and the value of T_c is not much affected. T_{c-zero} value of pure YBCO shows the highest temperature of 90 K while YBCO-15% Ag nano shows the lowest T_{c-zero}, 78 K. Although YBCO-15% nano Ag shows semiconductor behavior at normal state but it has a low resistance compared with these four other samples. This condition is caused by the presence of Ag in high quantity that have been improved the conductivity of the samples in normal state. Critical temperature, T_{c-zero} and change of critical temperature, ΔT_c ($T_{c-onset}$ - T_{c-zero}) and the behavior at normal state for the five samples have been summarized in Table 5.2.

Table 5.2. Critical temperature, T_{c-zero}, changes in critical temperature ΔT_c ($T_{c-onset}$ - T_{c-zero}) and normal state for five YBCO samples with weight percentage of nano Ag

Weight (%)	$T_{c-onset}$ (K)	T_{c-zero} (K)	ΔT_c (K)	Normal state
0.0	93	90	3	metal
2.0	93	85	8	metal
5.0	92	83	9	metal
10.0	94	84	10	metal
15.0	91	78	13	semiconductor

B) Phase Purity

The pure YBCO sample synthesized through ASG-SSR techniques with annealing temperature of 880 °C for 18 hours showed 123 single phase with lattice parameters a = 3.822 Å, b = 3.884 Å and c = 11.675 Å. From figure 5.4, found impurities peaks such as Y_2BaCu (Y-211), $BaCuO_2$ and CuO contents increases proportional with the percentage of nano Ag. The peaks indicated argentums in YBCO-15 % is clearly visible in the XRD pattern. However, the orthorhombic structure maintain for all samples. XRD data analysis found that the increasing of a lattice parameters and the reduction of c lattice parameter are proportional with the percentage of argentums respectively shown in figure 5.14. YBCO-2% and 5% nano Ag samples have the higher cells volume compared to pure YBCO. This condition is due to the entry of some Ag^+ ions into the orthorhombic structure and expensed its volume. However the orthorhombic structure volume contraction occurred in the YBCO-10% and 15% samples. This condition is due to the present of high quantity of nano Ag beyond the limit. The competition and the steric effects between Ag^+ ions from entering the cell structure is the reason to cause volume construction. This can be formulated that the presence of Y-211, $BaCuO_2$, as material side.

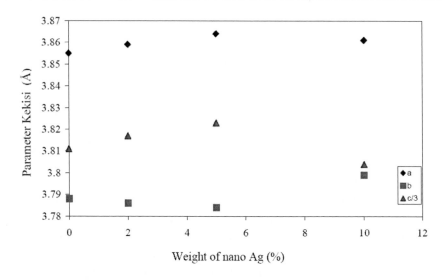

Figure 5.14. Lattice parameters *a, b* and *c* against percentage weight of nano Ag.

C) Critical Current Density, J_c

From Figure 5.5, J_c value of the samples increased with increasing the percentage of nano Ag. It is believed that Ag ions diffuse in to the grain boundary as metal and provides conductive paths between the superconducting grains. Consequently the amounts of weak links in the structure are reduced and the pinning capability is improved, which enhances the transport of critical current density in the samples. When the percentage of Ag is increased beyond a certain limit, the Y-211 phase increases and suppresses the superconducting behavior of $YBa_2Cu_3O_{7-\delta}$. This is illustrated by the 15% weight sample, which shows the lowest zero-resistance temperature of 78 K. Nevertheless, the improvement of J_c in the nano Ag added YBCO samples is due to the enhanced flux pinning, which is also observed in recent reports when metal Ag [Zhao et al. 2005; Gorur et al. 2005] were employed in the YBCO superconductors. From the experimental results the YBCO + 10% nano Ag has the highest J_c value of 2.3A/cm². From the XRD pattern result, when more and more percentage of nano Ag are added to the YBCO sample, foreign peaks due to Y_2BaCuO_5 (Y-211) and $BaCuO_2$ obtained rather than Y-123 phase. However, the existence of Y-211 are believed increase the value of critical current density, J_c. When the Ag contents in the sample increased beyond a certain limit, the presence of Y-211 phase also increased over Y-123 phase thus suppressed the superconductivity behavior [Salamati et al. 2001]. This situation occurs in the YBCO-15% where the value of $T_{\text{c-zero}}$ suppressed to 78 K from 90 K for pure YBCO. The result of J_c values at a relatively low for this nano-sized argentums additional YBCO sample compared with previous studies that use the micro-sized Argentums [Joo et al. 1998], Ag_2O [Abdelhadi et al. 1993] and Ag_2O_2 [Sato et al. 2002]. This situation is due to the presence of impurities in the YBCO-nano Ag samples such as Y_2BaCuO_5, $BaCuO_2$ and CuO have weakened the connection between the grain particles.

D) Micrograph of the YBCO + n% Nano Ag Sample

Argentums used in this work have uniform nano size show in Figure 5.6(a) which will get advantage in enhance interconnections between the grains. When nano Ag are added, the particles are uniformly distributed in the sample as shown in Figure 5.6 (b). When the amounts of Ag were increased to 10% weight, the nano Ag tends to agglomerate. The grain size also increased due to the melting and infiltrating of Ag particles, which were subsequently deposited between the grain boundaries of the structure. The YBCO-10% Ag sample shows a highly layered type structure with stronger link compared to the rest. This is an indication of much better links between the superconducting grains due to nano size Ag. The distribution of nano Ag as seen in Figure 5.6 (c) and 5.6 (d) causes a significant grain growth and partial melting of the system.

5.4.2. System $Y_{0.9}Ca_{0.1}Ba_{1.8}Sr_{0.2}Cu_3O_{7-\delta}$

In this work a bulk superconductor $Y_{0.9}Ca_{0.1}Ba_{1.8}Sr_{0.2}Cu_3O_{7-\delta}$ was successfully synthesized through solid state reaction between sol-gel precursor Ca-Ba-Sr-Cu-O and Y_2O_3 at 880 °C for 18 hour without post-anneal in oxygen. The sol-gel precursor Ca-Ba-Sr-Cu-O was successfully prepared by conventional metal oxide sol-gel method by using metal acetate as starting material. This sol-gel process involving hydrolysis-condensation of metal-organic compounds and the final product is polymerization metal-bond intramolecular -Ca-Ba-Sr-Cu-O-. The sol-gel precursor with apparent of organic material such as hydroxyl group –OH is nucleophile species which believe are chemically reactive to Y_2O_3 during solid-state reaction. Conventional solid-state reaction is considerably easy, especially if one or more of the starting materials are chemically reactive and contains ions that can diffuse easily. This novel route, ASG-SSR will not only enhances atomic mixing of reactants, the oxygen content for superconductor will also increase.

Many researchers [Fisher et al. 1993; Gasumyants et al. 2000; Jirak et al. 1998, Lin et al. 2002; Palles et al. 1998] have long attempted the substitution of Ca for Y in Y-123. Ca-doped YBCO is particularly of interest in the study of the electronic properties of CuO_2 planes and the C-O chains. Ca^{2+} valence is smaller than the one of the substituted Y^{3+} that can result in increasing number of hole [Awana et al. 1996; Chen et at. 1998; Jirak et al. 1998; Lin et al. 2002; Palles et al. 1998]. Anyway, increasing Ca content to replace Y in Y-123 phase was found to be accompanied by decreasing oxygen content [Fujihara et al. 1997; Lin et al. 2002]. According to Awana et.al (1996) on replacement of Y by Ca, the oxygen content of $Y_{1-x}Ca_xBa_2Cu_3O_{7-y}$ (y $\approx$0.3) system remains nearly unchanged, till x=0.1 and drops sharply to x=0.15, which is followed by decreasing bond distances Cu-O planes with increasing Cu(2)-O(2)-Cu(2) angle. Oxygen content or hole concentration (n_h) of Ca-doped $Y_2Ba_4Cu_7O_{14-y}$ phase was found to decrease with increasing Ca^{2+} content and reduce the superconducting transition temperature as indicated by Chen et al. (1998). In contrast, Ca-doped $YBa_2Cu_4O_{8-y}$ phase was found that increase the T_c. The first Ca-doped Y-124 phase was reported by Miyatake et al. (1989) with enhanced T_c to 91K. Hijar et al. (2001) reported that when Y was substituted by 2% of Ca , the T_c of Y-124 was found to be 87K compared to 80K of the undoped phase.

The studies on Sr-doped Ba site in Y-123 phase were conducted by several group [Ying et al.2001; Veal et al. 1987] in the aspect of oxygen order-disorder explanation. Sr substitution for Ba is expected to suppress the oxygen mobility due to the contraction of the unit cell and consequent compression of the Cu(1)-type sites[Veal et al. 1987], so T_c was reduced dramatically. Superconducting transition temperature, T_c for $Y(Ba_{1-x}Sr_x)_2Cu_3O_{7-y}$ with x= 0, 0.1, 0.2, 0.3, 0.4, and 0.6 were 91K, 85K, 83K, 80K and 78K, respectively [Ying et al. 2001]. Bael et al. (1998) reported that Sr^{2+} substitutes preferably in the Ba site but not in Y site in $Y(Ba_{1-x}Sr_x)_2Cu_4O_{8-y}$ phase and the critical temperature are increased to 88 K at 20 % Sr substitution. So far, no work on the double doping of Ca and Sr in Y-123 phase was reported in the literature. In this present work, synthesis of $Y_{0.9}Ca_{0.1}Ba_{1.8}Sr_{0.2}Cu_3O_{7-\delta}$ phase was conducted by novel method, which combining the sol-gel and solid state reactions.

A) Sol-Gel Ca-Ba-Sr-Cu-O Powder Characterization

An FTIR results for Ba-Sr-Ca-Cu-O multicomponent powder has been shown in Figure 5.8. Results show that the existence of the main peaks indicated -OH, -M-O-M-, absorption of H_2O and -COOH group.

B) *The Nature of Superconductivity*

Figure 5.10 shows the comparisons for the temperature dependence of resistance between the prepared bulk superconductor $Y_{0.9}Ca_{0.1}Ba_{1.8}Sr_{0.2}Cu_3O_{7-\delta}$ and pure YBCO. This decrement was well explained by the lattice distortion and deficiency of oxygen content. The lattice distortion is contributed by the difference of ionic radii between the doping element with the native element. Many reports [Awana et al. 1996; Chen et at. 1998; Fujihara et al.1997; Gasumyants et al. 2000; Jirak et al. 1998; Lin et al. 2002; Palles et al. 1998] show that with fixed oxygen stoichiometry, the substitution of Ca for Y generates holes at CuO_2 planes which has influence over superconducting transition temperature without changing the crystal structure of the sample due to slightly difference between the ionic radii. However, it was found that oxygen vacancies (which fill holes) are always co-introduced with Ca, thus reducing its holes generation effect [Fujihara et al.1997]. It is generally accepted that $T_{c\text{-zero}}$ is essentially determined by the holes concentration on the copper-oxygen planes (CuO_2) and the relative charge on oxygen within the planes. Meanwhile, substitution of smaller Sr ions for Ba ions bring about the lattice distortion for the crystal structure of this bulk superconductor. Such contraction of unit cell would suppress the oxygen mobility and follow by the decrease of $T_{c\text{-zero}}$ [Ying et al.2001, Veal et al. 1987]. Doping of Ca^{2+} ions into Y^{3+} layer will cause the total valence of the layer less than +3. This will bring about electron density flow into the CuO_2 layer decrease [Ying et al. 2001]. Although both samples showed metallic properties in normal conditions, but $Y_{0.9}Ca_{0.1}Ba_{1.8}Sr_{0.2}Cu_3O_{7-\delta}$ shows a low resistance at room temperature compared to pure YBCO. This result shows the present of Ca^{2+} and Sr^{2+} increased conductivity before the transition temperatures.

C) *Phase Purity*

Figure 5.11 show the XRD pattern for these both $Y_{0.9}Ca_{0.1}Ba_{1.8}Sr_{0.2}Cu_3O_{7-\delta}$ and pure YBCO. All the samples showed major XRD peaks can be indexed by a orthorhombic unit cell. Single phase compound was obtained for pure YBCO sample. Minor impurity peaks were observed in Ca and Sr doped YBCO sample, probably due to the semiconductor Y_2BaCuO_5 (Y-211) and $BaCuO_2$ phase, anyway it does not much influence the orthorhombic structure and superconducting properties. The lattice parameter and volume fraction for the nominal samples are listed in table 5.3. From XRD patterns, we conclude that YBCO phase can be stabilized under appropriate annealing condition prepared by this ASG-SSR.

**Table 5.3. Lattice parameter and volume fraction for YBCO
and $Y_{0.9}Ca_{0.1}Ba_{1.8}Sr_{0.2}Cu_3O_{7-\delta}$ annealing at 880 °C**

Sampel	a (Å)	b (Å)	c (Å)	%123	%211	%$BaCO_2$
$YBa_2Cu_3O_7$	3.824	3.886	11.681	100	0	0
$Y_{0.9}Ca_{0.1}Ba_{1.8}Sr_{0.2}Cu_3O_\delta$	3.887	3.829	11.690	66.32	20	13.68

D) SEM and EDAX

Figure 5.13 show that EDAX and percentage of atomics in bulk superconductor $Y_{0.9}Ca_{0.1}Ba_{1.8}Sr_{0.2}Cu_3O_{7-\delta}$. Excellence stoichiometry compound have been obtained for this bulk sample. We found that the average oxygen content of this bulk superconductor $Y_{0.9}Ca_{0.1}Ba_{1.8}Sr_{0.2}Cu_3O_{7-\delta}$ was determined to be 6.81 where $\delta= 0.19$. Although the whole heat treatment was done without oxygen flow, the product, $Y_{0.9}Ca_{0.1}Ba_{1.8}Sr_{0.2}Cu_3O_{6.81}$ was still in a highly oxidized state. It can believe that nucleophile species of hydroxyl group –OH in the sol-gel Ca-Ba-Sr-Cu-O precursor which is chemically reactive to Y_2O_3 during solid-state reaction has contribute extra oxygen and thus provides sufficient amount of oxygen for the sample to stabilized the polycrystalline.

YBCO sample showed fine, sharp grains 10 μm in size with no evidence of melting. The size of grain and porosity of sample decrease and these grain were agglomerate for the bulk $Y_{0.9}Ca_{0.1}Ba_{1.8}Sr_{0.2}Cu_3O_{7-\delta}$. Evidence of melting can be observed in the SEM. It can conclude that low temperature are required for this ASG-SSR route for preparing the YBCO superconductor.

REFERENCES

Abdelhadi, M.M. & Zip, K.A. 1993. The behavior of the flux flow resistance in YBCO/$(Ag_2O)_x$. Superconductor Science and Technology 7: 99-102.

Awana, V.P.S., Malik, S.K., Yelon, W.B. 1996. *Physica C* 262: 272.

Baranauskas, A., Jasaitis, D., Kareiva, A., Haberkorn, R. & Beck, H.P. 2001. Sol-gel preparation and characterization of manganese-substituted superconducting $YBa_2(Cu_{1-x}Mn_x)_4O_8$ compounds. *Journal of the European Ceramic Society* 21 : 399-408.

Chen, T.N., Yarng, S.L., Lin, J.P. 1998. *Chinese J. Phys.* 36.

Chuang, F. Y., Sue, D. J. and Sun, C. Y. 1995. *Mater. Res. Bull.* 30(10), 1300.

Costa, C. A and Mele, P. 2002. *Physica C.* 1174: 372-376.

Fan, Z. G., Shan,Y. Q., Wang,W. H., Wang, X. Y., Soh, D. W. and Zhao, Z. X., 1997. Physica C 495: 282 -287.

Fisher, B., Genossar, J., Kuper, C.G., Patlagan, L., Reisner, G.M., Knizhnik, A. 1993. *Phys. Rev. B.* 47: 6054.

Fujihara, S., Yoshida, N., Kimura, T., 1997. *Physica C.* 69:276.

Fujihara, S., Yoshida, N., Kimura, T., 1997. *Physica C.* 288: 158.

Gasumyants, V.E., Elizarova, M.V., Vladimirskaya, E.V., Patrina, I.B. 2000. *Physica C* 585: 341–348.

Görűr, O., Terzioglu, C., Varilci, A. and Altunbas, M. 2005. *Supercond. Sci. and Technol.* 18: 1233

Jirak, Z., Hejtmanek, J., Pollert, E., Triska, A., Vasek, P. 1998. *Physica C.* 165: 750.

Joo, J., Jung, S. B., Nah, W., Kim, J. Y. & Kim, T. S. 1999. Effects of silver additionals on the mechanical properties and resistance to thermal shock of $YBa_2Cu_3O_{7-\delta}$ superconductors. *Cryogenics* 39 :107-113.

Joo, J., Kim, J.G, and Nah, W. 1998. *Supercond. Sci. Technol.* 11: 645.

Kakihana, M. 1996. Sol-gel preparation of high temperature superconducting oxides. *Journal of Sol-Gel Science and Technology* 6: 7-55.

Karppinen, M., Kareiva, A., Linden, J., Lippmaa, M. and Niinistow, L. 1995. *J. Alloys and Compounds* 225: 586.

Lin, C.T., Liang, B., Chen, H.C. 2002. *J. Cryst. Growth* 778: 237–239.

Miyatake, T., Gotoh, S., Koshizuka, N., Tanaka, S. 1989. *Nature* 41:341.

Mathur, S., Shen, H., Lecerf, N., Filavi, M. H., Cauniene, V., Jorgensen, J. E. and A. Kareiva. 2002. J. *Sol-Gel Science and Technology.* 24: 57.

Orlavo, T. S., Laval, J. Y., Dubon, A.C., Nguyen-van-Huong, B. I., Smirnov and Y. P. Stepanov. 1998. *Superconducting Science and Technology.* 11: 467.

Palles, D., Liarokapis, E., Leventonri, T.H., Chakoumakos, B.C. 1998. J. *Phys.Condens. Matter.* 10: 2515.

Sato, T., Nakane, H., Yamazaki, S. & Mori, N. 2002. Analysis of fluctuation conductivity in melt-textured $YBa_2Cu_3O_{7-\delta}$ superconductor with Ag-doping. *Physica C* 372-376: 1208-1211.

Tepe, M., Avci, I., Kocoglu, H, and Abukay, D. 2004. *Solid State Communication* 131: 319

Veal, B.W., Kwok, W.K., Umezawa, A., Crabtree, G.W., Jorgensen, J.D., Downey, J.W., Nowicki, L.J., Mitchell, A.W., Paulikas, A.P., Sowers, C.H. 1987. *Appl. Phys. Lett.* 51: 279.

Van Beal, M.K., Kareiva, A., Vanhoyland, G., D'Haen, J., D'Olieslaeger, M., Franco, D., Quaeyhaegens, C., Yperman, J., Mullens, J., Van Poucke, L.C. 1998. *Physica C.* 307 :209.

Ying, X. N., Li, A., Huang, Y, N., Li, B. Q., Shen, H. M. & Wang, Y. N. 2001. The effect of strain on the low-temperature internal friction of Y(Ba$_{1-x}$Sr$_x$)$_2$Cu$_3$O$_{7-\delta}$. Journal of Physics: Condense Matter 13: 9813-9819.

Zhao, Y., Cheng, C. H and Wang, J. S. 2005. *Supercond. Sci. and Technol.* 18: S34

Zheng, X.G., Suzuki, M., Xu, C., Kuriyaki, H., Hirakawa, K., 1996. *Physica c* 271: 272.

Chapter 6

RESULTS AND DISCUSSION OF DATA ANALYSIS ON SUPERCONDUCTING SYSTEM RUTHENIUM

INTRODUCTION

This section discusses the results obtained from experiments carried out for the ruthenium copper oxide base system $RuSr_2GdCu_2O_{8-\delta}$ (Ru-1212), $RuSr_2(Gd,Ce)Cu_2O_{8-\delta}$ (Ru-1222) and $RuSr_{1.5}Ca_{0.5}PbCu_2O_{8-\delta}$ (RuPb-1212). Data obtained directly through the infrared spectrum (FTIR) of the sol-gel powder, four-point probe method for electrical resistance against temperature test (R-T), X-ray diffraction (XRD), SEM and EDAX further carried out on samples of superconductors to obtain complete results data. In the end of the chapter, discussion on the analysis of experimental results has been carried out in accordance with their respective series.

6.2. RESULTS AND ANALYSIS FOR $RuSr_2GdCu_2O_{8-\delta}$ (Ru-1212)

6.2.1. Infrared Spectrum of Sol-Gel Powder Ru-Sr-Gd-Cu-O

Infrared spectrum test was performed on the Ru-Sr-Gd-Cu-O sol-gel powder prepared by ASG shows in Figure 6.1. As FTIR results shown in Figure 4.13 and 5.1, there are four main regions. Peaks, 611.18 cm^{-1} and

670.21 cm^{-1} indicate the existence of M-O bond. The presence of –CO group is shown by the main peak of 1422.13 cm^{-1} and the fingerprint 680.66 cm^{-1}. Peak at 1551.52 cm^{-1} indicated –COOH group. Absorption width in the range 3365.37 cm^{-1} and the fingerprint on the peak 1049.11 cm^{-1} and 1021.16 cm^{-1} indicates the presence of –OH functional group. Water absorption is shown in the range of 3100-3700 cm^{-1}. The presence of -C≡C- functional groups are shown by the presence of medium intensity peak at 2328 cm^{-1} region.

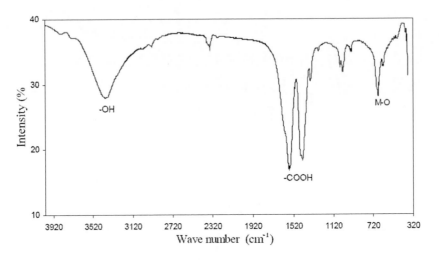

Figure 6.1. Infrared spectra of the Ru-Sr-Gd-Cu-O sol-gel powder prepared by the ASG technique.

6.2.2. Electrical Resistance Change against Temperature for Superconducting RuSr$_2$GdCu$_2$O$_{8-\delta}$ (Ru-1212)

The resistances versus temperature curve for the Ru-1212 annealed at 950 °C, 1000 °C and 1030 °C for 24 hours are shown in Figure 6.2. The figure shows that low annealing temperature is not sufficient to form the superconducting phase. The sample annealed at 950 °C shows insulator like normal state behavior $T_{\text{c-onset}}$ 30 K but no zero resistance temperature ($T_{\text{c-zero}}$). For 1000 °C annealing temperature, the sample showed insulator–semiconductor-like behavior in the normal state. The resistance dropped sharply with $T_{\text{c-onset}}$ at 43 K and $T_{\text{c-zero}}$ close to 37 K. The sample annealed at 1030 °C showed semiconductor-like normal state behavior and room temperature resistivity in the order of 10^{-4} Ω. This temperature is the optimum annealing temperature for superconductivity in this study with superconducting transition temperature, $T_{c\text{-onset}}$ near 55 K and $T_{c\text{-zero}}$ at 45 K. The upturn of resistance near 140 K as the temperature is lowered has been ascribed to a magnetic transition [Awana 2005].

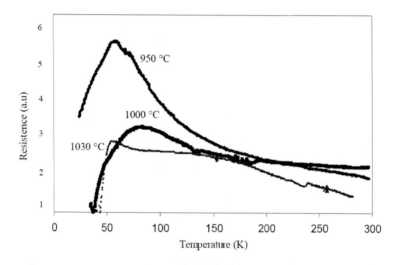

Figure 6.2. Electrical resistances against temperature for Ru-1212 at annealing temperature 950 °C, 1000 °C and 1030 °C for 24 hours.

6.2.3. X-Ray Diffraction Patterns for RuSr$_2$GdCu$_2$O$_{8 \text{-} \delta}$ (Ru-1212)

Figure 6.3 shows the XRD pattern for the sample annealed at 1030 °C. Impurity peaks in the XRD patterns decrease with an increase in the annealing temperature. The lattice parameters for the sample annealed at 1030 °C are $a=b=3.832$ Å, $c=11.478$ Å and the peaks can be assigned to the space group I4/mmm. From the XRD patterns, a nearly single phase RuSr$_2$GdCu$_2$O$_{8\text{-}\delta}$ compound was obtained with a small amount of SrRuO$_3$. A total of 95.65% of the XRD peaks, is contributing to the Ru-1212 phase and around 4.35 % of peaks are attribute to RuSrO$_3$.

Pure Ru-1212 has been successfully synthesized through the acrylamide polymerization sol–gel method but is not superconducting [Zhigadlo et al. 2003] It is interesting to note that Ru-1212 was found to be non-superconducting when no trace of SrRuO$_3$ is present [Awana 2005].

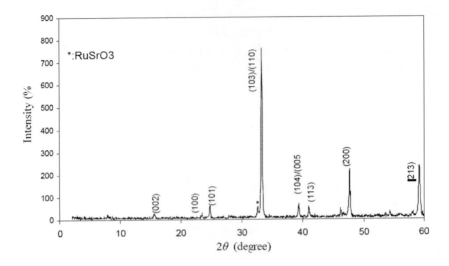

Figure 6.3. The XRD diffraction spectrum (XRD) of $RuSr_2GdCu_2O_{8-\delta}$ annealed at 1030°C for 24 hours.

6.2.4. Scanning Electron Micrographs (SEM) for Ru-1212

The scanning electron micrographs for the Ru-1212 annealed at 1030 °C shown in Figure 6.4 indicate a fine, sharp grain 1–5 μm in size with no evidence of melting.

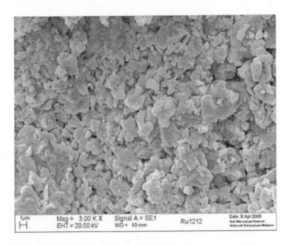

Figure 6.4. SEM micrographs of Ru-1212 samples.

6.3. RESULTS AND ANALYSIS FOR RuSr₂(Gd,Ce)Cu₂O₁₀-δ (RU-1222)

6.3.1. Infrared Spectrum (FTIR) of Ru-Sr-Gd-Ce-Cu-O Sol-gel powder

FTIR test was performed on the Ru-Sr-Gd-Cu-O sol-gel powder prepared by ASG shows in Figure 6.5. There are four main regions in the curve graph which have qualitatively similar in Figure 6.1.

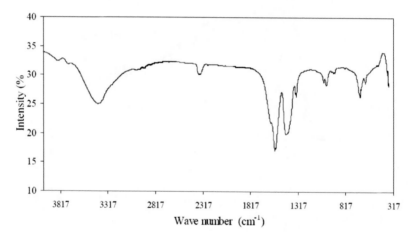

Figure 6.5. Infrared spectra of Ru-Sr-Gd-Ce-Cu-O the sol-gel powder.

6.3.2. Electrical Resistance Change against Temperature for Ru-1222 Superconducting

The resistance against temperature curves for $RuSr_2(Gd_{2-x}Ce_x)Cu_2O_{10-\delta}$ series with $x = 0.4, 0.5, 0.6, 0.7$ and 0.8 with 1050 °C annealing temperature for 24 hours are shown in Figure 6.6. All samples showed semiconductor-like normal state behavior. The highest onset transition temperature, $T_{c\text{-onset}}$, was 59 K. A zero-resistance temperature, $T_{c\text{-zero}}$, of 40 K was observed in the $x = 0.6$ sample. This transition temperature is slightly higher than samples where Gd was replaced by Eu and Sm, prepared by the conventional solid state reaction method [Balchev et al. 2005]. The $x = 0.5$ and 0.7 samples show $T_{c\text{-zero}}$ of 30 K and 16 K, respectively. The impurities may have caused a suppression of the transition temperature in these samples. Among the five samples prepared, only the $x = 0.5$, 0.6 and 0.7 which shows the nature of superconductivity. All samples show semiconducting properties as the normal state before the temperature of 48 K. For samples with $x = 0.4$ and 0.8 show $T_{c-onset}$ 41 K and 38 K respectively, but did not reach the zero critical temperature.

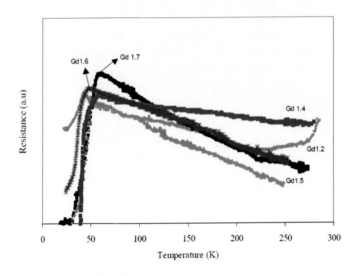

Figure 6.6. Electrical resistance against temperature curve of $RuSr_2(Gd_{2-x}Ce_x)Cu_2O_{10-\delta}$ with $x = 0.4$, 0.5, 0.6, 0.7, 0.8 series.

6.3.3. X-Ray Diffraction Pattern for Ru-1222

Figure 6.7 shows the XRD pattern for the $RuSr_2(Gd_{2-x}Ce_x)Cu_2O_{10-\delta}$ samples, (x= 0.5, 0.6 and 0.7). The diffracted pattern can be indexed to a tetragonal unit cell with space group I4/mmm. From the XRD pattern, a $SrRuO_3$ impurity was observed in the $x = 0.5$ sample. CeO_2 was observed in the sample with $x = 0.7$, probably due to the fact that a sample prepared this way has exceeded the Ce solubility limit. A single phase compound with no impurity was obtained for the $x = 0.6$ sample, with tetragonal symmetry and lattice parameters $a = b = 3.863$ Å and $c = 28.57$ Å. In contrast, samples prepared by the conventional solid state reaction method normally showed extra peaks, which were attributed to impurities such as $(Sr_{1-x}Gd_x)(Ru_{1-y}Cu_y)O_{3-d}Gd_2-CuO_4$ [Matveev et.al 2005]. Table 6.1 shows that all the samples have a different percentage of foreign phase according to the value of x.

Sample $x = 0.4$ has minimum 1222 phase ($\sim 40\%$) while the $x = 0.6$ has the highest 1222 phase ($\sim 97\%$).

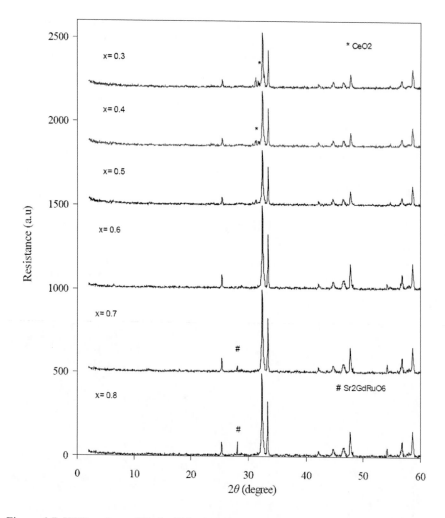

Figure 6.7. XRD pattern of $RuSr_2(Gd_{2-x}Ce_x)Cu_2O_{10-\delta}$ samples, with $x= 0.3, 0.4, 0.5, 0.6,$ 0.7 and 0.8.

Table 6.1. $T_{c\text{-onset}}$, $T_{c\text{-zero}}$, **percentage of 1222 phase and foreign phase and lattice parameters of the Ru-1222**

x	$T_{c\text{-start}}$ (K)	$T_{c\text{-zero}}$ (K)	% 1222	% CeO_2	% Sr_2RuGdO_6	a (Å) (± 0.01)	c (Å) (± 0.01)
0.4	41	-	40	27	33	3847	28.85
0.5	58	30	75	15	10	3842	28.59
0.6	59	40	90	7	3	3863	28.57
0.7	49	16	66	23	11	3839	28.54
0.8	38	-	66	18	16	3837	28.53

Lattice parameter a increases and the lattice parameter c decrease is equivalent to increasing of Ce doping in the sample. This condition is due to the size of Ce^{4+} has smaller radius compared to Gd^{3+}.

6.3.4. Scanning Electron Micrographs (SEM) for Ru-1222

The scanning electron micrograph for $x = 0.5$ and 0.6 is shown in Figure 6.8. Fine, sharp grains with size 1-5 μm, with evidence of partial melting, are observed in all samples.

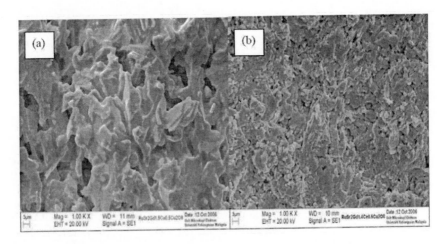

Figure 6.8. SEM micrographs for (a) $RuSr_2Gd_{1.5}Ce_{0.5}Cu_2O_{10-\delta}$ dan (b) $RuSr_2Gd_{1.4}Ce_{0.6}Cu_2O_{10-\delta}$.

6.4. RESULTS AND ANALYSIS
OF RuSr$_{1.5}$CA$_{0.5}$PBCU$_2$O$_{8-\delta}$ (RUPB-1212)

6.4.1. Infrared Spectrum of Sol-Gel Powder
Ru-Sr -Ca-Pb-Cu-O

FTIR test was performed on the Ru-Sr-Gd-Cu-O sol-gel powder prepared by ASG shows in Figure 6.5. There are four main regions which are refer to H$_2$O absorption, -COOH, -MO and –OH groups in the curve graph and have similarity in Figure 6.1and 6.5. The quality of the gel powder are remain the same and ready for use.

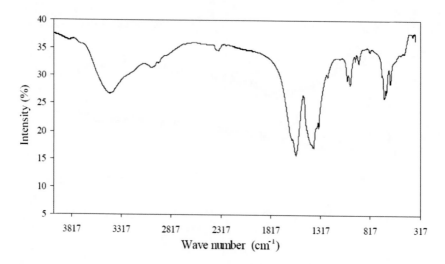

Figure 6.9. Infrared spectra of the sol-gel powder, Ru-Sr-Ca-Pb-Cu-O prepared ASG technique.

6.4.2. Electrical Resistance against Temperature for Superconducting RuPb-1212

Figure 6.10 show the electrical resistance against temperature curves for RuPb-1212 annealing at 850 °C, 890 °C and 920 °C for 24 hours. The sample annealed at 850°C shows insulator-like normal state behavior starting at 90 K, and with $T_{c\text{-onset}}$ at 20 K and $T_{c\text{-zero}}$ close to 10 K. The sample annealed at 890 °C showed metallic-like normal state behavior with a slight upturn near 70 K and below. The room temperature resistivity is in the order of 10^{-4} μm. The 890 °C temperature is the optimum annealing temperature where the sample exhibits an onset transition temperature, $T_{c\text{-onset}}$ at 35 K and $T_{c\text{-zero}}$ at 20 K. Insulator-like normal state behavior was observed when the annealing temperature was increased to 920 °C and the nature of superconductivity RuPb-1212 sample has been destroyed.

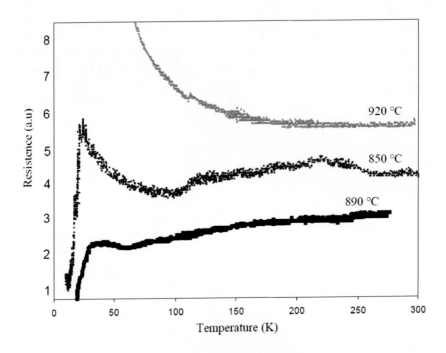

Figure 6.10. The electrical resistance versus temperature curve of $RuSr_{1.5}Ca_{0.5}PbCu_2O_{8-\delta}$ annealed at 850 °C, 890 °C and 920 °C.

6.4.3. X-Ray Diffraction Patterns for RuPb-1212

Figure 6.11 shows the XRD patterns for samples annealed at 850°C, 890°C and 920°C for 24 hours. The XRD patterns showed that all samples exhibited a dominant 1212 type phase and can be indexed to a tetragonal unit cell with space group I4/mmm. A pure phase (=97%) of 1212 type was successfully synthesized through annealing at 890°C, without the presence of $SrRuO_3$. A nearly single-phase compound was obtained with a small amount of $SrRuO_3$ in samples annealed at 850°C and 920°C. The lattice parameters and unit cell volume are larger than the typical $RuSr_2GdCu_2O_{8-\delta}$ and are ascribed as due to the different elements involved in the materials studied. Table 2 has shown that phase in 1212 and $SrRuO_3$ and lattice parameter of the sample by different annealing temperature.

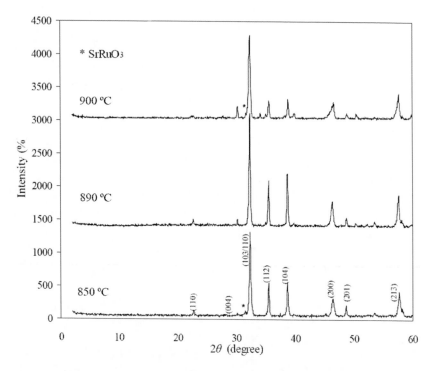

Figure 6.11. The XRD diffraction pattern of $RuSr_{1.5}Ca_{0.5}PbCu_2O_{8-\delta}$ annealed at (a) 850 °C, (b) 890 °C and (c) 920°C for 24 hours (* is the impurity peak $SrRuO_3$).

Tetragonal unit cell volume of samples increased with increasing annealing temperature. This is because the increase in treatment temperature has expended the radius size of cations. The presence of $SrRuO_3$ peak is most obvious for the samples treated with annealing temperature of 850 °C showing the ratio of 5.95 for the peaks 103/110.   Samples treated with annealing temperature of 890 °C showed the presence of minimum $SrRuO_3$ peak with only about 3% compared with the rest.

Table 6.2. $T_{c\text{-onset}}$, $T_{c\text{-zero}}$, **percentage of 1212 phase, foreigner phase and lattice parameters *a, b* and *c* for RuPb-1212**

Annealing Temperature	$T_{c\text{-onset}}$ (K)	$T_{c\text{-zero}}$ (K)	% 1212	% SrRuO$_3$	a (Å) (± 0.01)	c (Å) (± 0.01)
850 °C	21	10	95	5	3.91	11.74
890 °C	30	20	> 97	< 3	3.92	11.76
920 °C	-	-	< 95	> 5	3.92	11.81

6.4.4. Scanning Electron Micrographs (SEM) for RuPb-1212

Figure 6.12 shows the results of SEM micrographs of the expansion of KX 5.00 carried out on RuPb-1212 samples with annealing temperature of 850 °C, 890 °C and 920 °C for 24 hours. Sample annealed at 850°C showed fine, sharp grains with size of 1–5 μm and some evidence of melting. The grain size and porosity of the sample decreased when the sample was annealed at 890°C. The grains were agglomerate when the annealing temperature was increased to 920 °C. there have evidence of melting observed from the increase the grain size to ~10μm. RuPb-1212 are not resist to high annealing temperature due to the present of lead element on the sample and lower it melting point.

Figure 6.12. Continued on next page

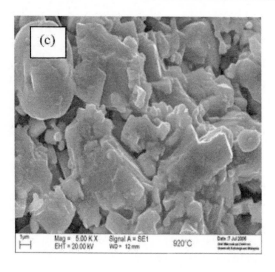

Figure 6.12. SEM micrographs of $RuSr_{1.5}Ca_{0.5}PbCu_2O_{8-\delta}$ prepared by the ASG sol–gel route annealed at (a) 850 °C (b) 890 °C and (c) 920 °C.

6.5. DISCUSSION

Ruthenocuprates system, $RuSr_2GdCu_2O_{8-\delta}$ (Ru-1212) and $RuSr_2(Gd_{2-x}Ce_x)Cu_2O_{10-\delta}$ (Ru-1222) are extraordinary example of high-T_c cuprates because of co-existence with the nature of superconductivity and ferromagnetic properties [Felner et.al 2000]. Both Ru-1212 and Ru-1222 phase are structurally related to $YBa_2Cu_3O_{7-\delta}$ (YBCO), phase with Cu in charge reservoir replaced by Ru such that Cu-O chain is replace by RuO_2 sheet. For Ru-1222 structure, a three layer fluorite-type block $-[(Gd,Ce)-O]_3$- instead of a single oxygen free R (rare earth element) layer is inserted between the two CuO_2 planes of the YBCO [Awana et.al 2003]. The replacement of Cu in the charge reservoir block by the higher-valence increases the overall oxygen content. It is believe that CuO_2 planes is responsible for the natural of superconductivity while ferromagnetic ordering take place in $RuSrO_3$ block [Petrykin et al. 2002]. Both Ru-1212 and Ru-1222 are belong to p-type superconducting (hole type superconductor). In the case of Ru-1222 system, Ce^{4+} doped into the layer of Gd^{3+} thus total valance in the fluorite-type block become higher than +3. Therefore, excess electrons will enter into the CuO_2 layers to balance the overall charge.

6.5.1. RuSr$_2$GdCu$_2$O$_{8-\delta}$ (Ru-1212Gd) Superconductor

The RuSr$_2$GdCu$_2$O$_{8-\delta}$ (Ru-1212) which belongs to the cuprate family of high temperature superconductor (HTSC) which was first synthesized in 1995 [Bauernfeind et al. 1995] has seen a strong ongoing interest among scientists in the field. Superconductivity (SC) and ferromagnetism (FM) are two antagonistic phenomena but were found to coexist in this layered RuSr$_2$GdCu$_2$O$_{8-\delta}$ [Chmaissem et al. 2000; Tang et.al 1997].

The unusual behavior and characteristic of Ru-1212 have triggered much interest among researchers in the synthesis of this ferromagnetic superconductor. Technique preparation of superconducting Ru-1212Gd is a complicated and tricky knowledge because of the existence of Gd and Ru plays a role in breaking the Cooper pairs. The conventional solid-state method is the normal route to synthesize this material [Nachtrab et al.2004; Lorenz et al. 2001; Yokosawa et al. 2004]. The solid-state reaction involving high purity oxides and carbonate powders has several drawbacks. For example it requires extended heat treatment and repeated grinding in order to obtain a nearly single phase. The resultant materials were also non-homogeneous. High annealing temperature (1050 °C–1060 °C) and longer heating duration would bring about the decomposition of the Ru-1212 phase associated with evaporation of ruthenium oxide which consequently lowers the quality of the samples [Matveev et al. 2004]. The Ru-1212 ferromagnetic superconductor was first synthesized where an extra step of annealing in flowing nitrogen at 1010 °C was introduced to reduce the impurity phase, RuSrO3. It was finally annealed in flowing oxygen at 1050 °C for 72 h, in order to obtain a fully oxygenated Ru-1212 with superconducting transition at 16 K. The step consisting of annealing sample in flowing nitrogen caused too much oxygen out of the structure and is contradictory to the subsequent effort of putting more oxygen back into the structure [Matveev et al. 2004]. It indirectly lowers the oxygen content of the sample. The usual solid-state reaction is not the best method in spite of prolonged annealing in oxygen [Wang et al. 2003] and even in high oxygen pressure [Awana et al. 2002]. Sol–gel technique which involved acrylamide polymerization process was successful in synthesizing pure phase of Ru-1212 [Zhigadlo et al. 2003] and solid-state reaction at high oxygen pressure [Henn et al. 2000; Steiger et al. 2007] but does not indicate the nature of superconductivity. In this work, we synthesized superconducting Ru-1212 by using the sol–gel route. We found that this route is useful in obtaining almost single phase superconducting Ru-1212 with zero resistance temperature as high as 45 K. The sol–gel method is generally better that the solid-state reaction in obtaining chemical homogeneity and chemical reactivity. Moreover, lower heating temperature and shorter annealing duration can be obtained through the sol–gel route compared to that prepared by the traditional solid-state reaction.

$RuSr_2GdCu_2O_{8-\delta}$ (Ru-1212) superconductor was successfully synthesized through the ASG sol–gel route with $T_{c\text{-onset}}$ near 55 K and $T_{c\text{-zero}}$ at 45 K. Optimization of synthesis and annealing conditions may further improve the superconducting property. We found that 1030 °C is the optimum annealing temperature for the Ru-1212 synthesized by the sol–gel route. The crystal structure was found to consist of ~92% tetragonal phase in 1212 and ~8% $SrRuO_3$ peak for the sample synthesized through the optimun, 1030 °C. These results are similar to previous studies [Wang et al. 2002; Nachtrab et al. 2004]. Hence, the sol–gel route is an alternative way to prepare theRu-1212 superconductor.

6.5.2. $RuSr_2(Gd_{2-x}Ce_x)Cu_2O_{10-\delta}$ (Ru-1222) Superconductor

Bauernfeind et al. (1995) have shown that the parameters of Gd/Ce ratio determine the nature of conductivity in the normal state and superconductivity. The hole doping of the Cu-O planes can be optimized with an appropriate variation of Gd/Ce ratio [Balchev et.al 2005]. Thus the factor of homogeneity is very important in this multicomponent compound preparation. Solid-state reaction is usually used by scientist to synthesized this Ru-1222 were very tricky and not guarantee. Repeated grinding, high annealing temperature 1050 °C-1060 °C at 1 atm and long periods are needed for Ru-1222 synthesized. However, this technique is difficult to obtain a single phase due to high annealing temperature and long annealing duration bring about decomposition of the Ru-1222 phase via evaporation of ruthenium oxide, consequently lowers the quality of the samples [Matveev et al. 2004]. An oxygen atmosphere of 1.0 to 1.2 atm is required to obtain this phase [Balchev et al. 2005].

The ASG sol-gel method is an alternative method to obtain chemical homogeneity, as well as high purity, in a sample. Moreover, the sol-gel route does not require such high heating temperatures, and the annealing duration is short compared to that of the conventional solid-state reaction. According to previous research, the single-phase $RuSr_2Gd_{1.5}Ce_{0.5}Cu_2O_\delta$ compound was successfully synthesized using the polymer complex method. The oxygen partial pressure was controlled to prevent Ru loss in the sealed volume [Petrykin et al. 2002]. The sol-gel technique, involving an acryl amide polymerization process, was successful in synthesizing a pure phase of $RuSr_2GdCu_2O_8$ but was not superconducting.

In previous reports, it was shown that the $Ru(Sr_{1.5}Ca_{0.5})PbCu_2O_8$ superconductors could be synthesized by the sol-gel route [Yeoh et al. 2008]. In this study, we success to synthesis superconducting $RuSr_2(Gd_{2-x}Ce_x)Cu_2O_{10-\delta}$, with x =0.5, 0.6, 0.7, by ASG sol-gel method. This route resulted in the formation of single phase superconducting Ru-1222 for x = 0.6, in a one-step process. Critical temperature varies systematically with Ce content, as shown in Figure 6.6 and only samples with x = 0.5, 0.6, 0.7 shows the nature of superconductivity. T_{c-zero} for all three samples are 16 K-40 K with different Ce content. This result was somewhat comparable to the critical temperature of previous studies [Bauernfeind et al.1995; Felner et al. 1997 Williams et al. 2002 and Lorenz et al. 2001] but higher than the study of Tang et al. (1996) with $T_{c-onset}$ 30 K and T_{c-zero} 10 K. In addition, the $T_{c-onset}$ and T_{c-zero} of this study is higher than the Nb-1222 (27 K, 13 K) and Ta-1222 (30 K, 10 K) which is not ferromagnetic superconductor [Abd-Shukor et al. 2005].

All XRD peaks can be indexed with tetragonal unit cell except for the present of foreign peak objects due to Sr_2GdRuO_6, CeO_2 and CuO. Formation of Ru-1222 system structure is quite complicated, less stable and difficult synthesized. x = 0.6 sample has a maximum superconductivity shows that the most pure 1222 phase. In this system, nitrogen gas is used to prevent the formation of $SrRuO_3$, but will result in the formation of Sr_2GdRuO_6, CeO_2 and CuO. According to Bauernfeind et al. (1995), Tang et al. (1996) and Williams et al. (2000) reports, the presence of foreign material can be eliminated with prolong the calcinations temperature under oxygen flow for several days.

From the XRD data analysis, the a and c lattice parameters are gradually reduced with increasing of Ce doping content (with increasing the value of x). Since the radius of Gd^{3+} (1053 Å) is longer than the radius of Ce^{4+} (0.97 Å), the replacement of Ce in Gd will cause the size of the radius of the (Gd, Ce)<Gd, consequently shrinks the size of the tetragonal unit cell c axis. An axis is dependent on the size of the Cu-O framework. Besides that, the doping of Ce^{4+} on Gd^{3+} will cause the valance of Cu ions increase as well, therefore the lattice parameter a was reduced.

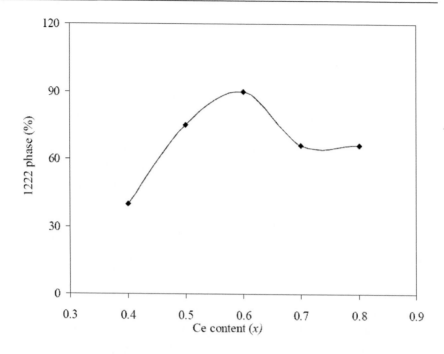

Figure 6.13. The percentage of 1222 phase versus x for $RuSr_2(Gd_{2-x}Ce_x)Cu_2O_{10-\delta}$

6.5.3. $RuSr_{1.5}Ca_{0.5}PbCuSr_2O_{8-\delta}$ (RuPb-1212)

According to Rose-Innes and Rhoderick (1994), copper oxide material is superconductivity properties of a substance that does not change significantly when the composition plane in the middle replaced. The accidentally discovery of ferromagnetic superconducting Ru-Cu systems are due to the idea that aims to improve the critical current density with the doping or insertion of the metal layer in high temperature copper oxide superconducting material. Because the plane of the RuO_2 layer has a long rectangular plane coordination bonding is similar to the CuO_2, then it is considered suitable for replacing CuO_2 in HTSC. The further study found that the Ru-Cu system is very unique because the display properties ferromagnetic before the Curie temperature superconductivity transition temperature [Felner et al. 1997; Lynn et al. 2000].

While searching for the mystery that exists in the ferromagnetic superconducting system of Ru-1212 and Ru-1222, various studies have been done to change the parts or components by doped with certain other metals. Lanthanide metal series have been found by scientists to replace Gd and Ru element in the system. Gd have strong magnetic properties and play the role of Cooper pairs breaking. Awana et al. (2001) successfully synthesized $RuSr_2PrCu_2O_{8-\delta}$ with the same tetragonal structure with Ru-1212 showing magnetic order at 150 K. In addition, the Sm was doped in $RuSr_2GdSm_{4.1}Ce_{0.6}Cu_2O_{10-\delta}$ and shows magnetic order at 130 K [Oomi et al. 2002]. $RuSr_2Eu_{2-x}Ce_xCu_2O_{10-\delta}$ shows a superconducting transition at 45 K [Awana et al. 2003].

Pb substitution has been employed in enhancing the formation cuprate based superconductors such as $(Bi,Pb)_2Sr_2Ca_2Cu_3O_{10-\delta}$ and $(Tl,Pb)Sr_2Ca_2Cu_3O_9$. Shi et al (2003) reported the effect of Pb-doping on the Ru site in $Ru_{1-x}Pb_xSr_2Gd_{1.4}Ce_{0.6}Cu_2O_{10-\delta}$. Unlike in the Bi and Tl-based compounds where phase formation and the transition temperature were enhanced, Pb-doping in the Ru-1222 phase suppresses the transition temperature. We report the preparation of Pb containing Ru-1212-type superconductor with starting formula $Ru(Sr_{1.5}Ca_{0.5})PbCu_2O_8$ through the sol–gel route. In this study, we found that complete elimination of the $SrRuO_3$ impurity gives the highest transition temperature. Pb together with Ca lowered the temperature of formation of the Ru-1212 type phase from 1000 °C in previous studies to 890 °C.

Three different condition with calcination temperatures of 850 °C, 890 °C dan 920 °C for 24 hours under oxygen flow were carried out on RuPb-1212 systems. Samples treat with low calcinations temperature, 850 °C and 890 °C show superconducting behavior. 890 °C is the optimum calcinations temperature for RuPb-1212 sample shows $T_{c\text{-onset}}$ 30 K and $T_{c\text{-zero}}$ 20K. Transition critical temperature of $T_{c\text{-onset,}}$ $T_{c\text{-zero}}$ and the normal state of RuPb-1212 samples synthesized under three different calcinations temperature treatments are listed in Table 6.3. It was found that high calcinations temperature, 920 °C are not suitable to stabilized RuPb-1212 phase formation. It is due to evaporation of Pb component occurring under high temperature treatment. The present of Pb and Ca lowered the temperature of formation of the Ru-1212 type phase. It is interesting to investigate further the various properties of this superconductor, especially the magnetic measurements.

Table 6.3. The critical temperature $T_{c\text{-onset}}$, $T_{c\text{-zero}}$ and normal state of RuPb-1212 annealing at 850 °C, 89 °C 920 °C

Calcination temperature (°C)	$T_{c\text{-onset}}$ (K)	$T_{c\text{-zero}}$ (K)	Normal state
850	21	10	Semimetal
890	30	20	metal
920	-	-	insulator

XRD pattern in Figure 6.12 shows all the peaks are contributing to the tetragonal phase in 1212. Samples treated with the temperature 890 ° C shows single phase RuPb-1212 in the absence of foreign objects such as $SrRuO_3$ but shows superconductivity behavior. This system is different compared with previous studies which emphasized that the presence of $SrRuO_3$ peak is necessary to show the nature of superconductivity. Figure 6.14 shows the phase purity of 1212 for all three samples RuPb-1212 with three different temperature treatments.

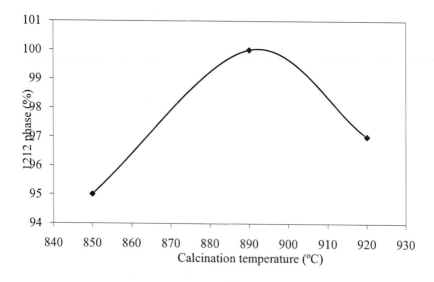

Figure 6.14. The percentage of 1212 phase versus temperature for sample treatment $RuSr_{1.5}Ca_{0.5}PbCu_2O_{8-\delta}$.

From the analysis of XRD data, the lenght of a lattice parameter increase while c lattice parameter decrease for RuPb-1212 compare to Ru-1212. The dopping of Ca^{2+} (0.99 Å) into Sr^{2+} (1.13 Å) will cause the contraction of size, $(Sr, Ca)_2 < Sr_2$ in the unicell of crystal structure. However, the present of Ca in small quantities does not much affect the length of lattice contraction and this may also due to the present of Pb^{2+} in the system. Since the radius of Pb^{2+} (1.20 Å) greater then radius of Gd^{3+} (1.02 Å), thus the replacement of Gd with Pb has expanded the size of the tetragonal unit cell c axis. The axis of a is depend on Cu^{3+}-O chain, the present of Pb^{2+} will lower the valence of Cu^{3+}, thus bring about the increasing of a lattice parameter in RuPb-1212 system.

REFERENCES

Abd-Sukor, R., Kek, W. K. & Yeoh, W., K. 2005. Superconducting properties of Ce-substituted $TaSr_2(Gd,Ce)_2Cu_2O_y$. *Journal of Materials Science*. 20: 50-56.

Awana, V. P. S. 2005. Magneto-superconductivity of rutheno-cuprates. A.V. Narlikar Edition, Frontiers in Magnetic Materials, Germany: p. 531. Springer-Verlag Publishing.

Awana, V. P. S., Ichihara, S., Karppinen, M. & Yamauchi, H. 2003. Comparison of magneto-superconductive properties of $RuSr_2RuCu_2O_{8-\delta}$ and $RuSr_2Gd_{1.5}Ce_{0.5}Cu_2O_{10-\delta}$. *Physica C* 378-381: 249-254.

Awana, V. P. S., Nakamura, J., Karppinen, M., Yamauchi, H., Malik, S, K. & Yelon, W. B. 2001. Synthesis and magnetism of Pr-based rutheno-cuprate compound $RuSr_2PrCu_2O_{8-\delta}$. *Physica C* 375-360: 121-125.

Balchev, N., Kunev, B., Pirov, J., Mihova, G. & Nenko, K. 2005. Low temperature magnetoresistance in Ru-1222 superconductor. *Materials Letters*. 59: 2357-2360.

Bauernfeind, L., Widder, W. & Braun, H.F. 1995. Ruthenium-based layered cuprates $RuSr_2LnCu_2O_8$ and $RuSr_2(Ln_{+x}Ce_{1-x})Cu_2O_{10}$ (Ln= Sm, Eu and Gd). *Physica C* 245:151-158.

Chmaissem, O., Jorgensen, J. D., Shaked, H., Dollar, P. & Tallon, J. L. 2000. Crystal and magnetic structure of ferromagnetic superconducting $RuSr_2$-$GdCu_2O_8$. *Phys. Rev.* B 61:6401-6407.

Felner, I., Asaf, U., Levi, Y. & Millo, O. 1997. Coexistence of magnetism and superconductivity in $R_{1.4}Ce_{0.6}RuSr_2Cu_2O_{10-\delta}$ (R = Eu and Gd) *Physical Review B* 55: R3374-R3377.

Lorenz, B., Meng, R. L., Cmaidalka, J., Wang, Y. S,. Lenzi, J., Xue, Y. Y. & Chu, C. W. 2001. Synthesis, characterization and physical properties of the superconducting ferromagnet $RuSr_2GdCu_2O_8$. *Physica C* 363: 251-259.

Lynn, J. W., Keimer, B., Ulrich, C., Bernhard, C. & Tallon, J. L. 2000. Antiferromagnetic ordering of Ru and Gd in superconducting $RuSr_2GdCu_2O_8$. *Physical Review* B 61(22): R14964-R14967.

Matveev, A. T., Sader, E., Duppel, V., Kulakov, A., Maljuk, A., Lin, C. T. & Habermeier, H. U. 2004. Decomposition of $RuSr_2GdCu_2O_8$ phase under high-temperature treatment. *Physica C* 403: 231-239.

Nachtrab, T., Koelle, D., Kleiner, R., Bernhard, C. & Lin, C. T. 2004. Intrinsic Josephson Effects in the magnetic superconductor $RuSr_2GdCu_2O_8$. *Physical Review Letters* 92(11): 117001-1-117001-4.

Oomi, G., Honda, F., Ohashi, M., Eto, T., Hai, D., Kamisawa, S., Watanabe, M. & kadowaki, K. 2002. Effect of pressure on the superconductivity of $RuSm_{1.4}Ce_{0.6}Sr_2Cu_2O_{10.}$ *Physica B* 312-313: 88-90.

Petrykin, V. V., Kakihana, M., Tanaka, Y., Yasuoka, H., Abe, M. & Eriksson, S. 2002. Ferromagnetic and superconducting properties of pure $RuSr_2Ce_{0.5}Eu_{1.5}Cu_2O_{10}$ samples prepared by polymerizable complex method. *Physica C* 378-381: 251-259.

Rose-Innes, A. C. & Rhoderick, E. H. 1994. Introduction to superconductivity. Edition ke-2. Oxford: Pergamon Press.

Shi, L., Li, G., Pu, Y., Zhang, X. D., Feng, S. J. & Li, X. G. 2003. Effect of Pb dopng on the superconducting and magnetic resonance properties of Ru-1222. *Materials Letters.* 57: 3919-3923.

Steiger, M., Kongmark, C., Rueckert, F., Harding, L. & Torikachvili, M. S. 2007. Pressure study of superconductivity and magnetism in pure and Rh-doped $RuSr_2GdCu_2O_8$ materials. *Physica C* 453: 24-30.

Tang, K. B., Qian, Y. T., Yang. L. Y., Zhao, D. & Zhang, Y. H. 1997. Crystal structure of a new series of 1212 type type cuprate $RuSr_2LnCu_2O_z$. *Physica C* 282-287: 947-948.

Tang, K. B., Qian, Y. T., Zhao, Y. D,. Yang. L., Chen, Z. Y. & Zhang, Y. H. 1996. Synthesis and characterization of a new layered superconducting cuprate: $RuSr_2(Ce,Gd)_2Cu_2O_z$ *Physica C* 259: 168-172.

Wang, D. Z., Ha, H. L., Oh, J. L., Moser, J., Wen, J. G., Naughton, M. J. & Ren, Z. F., 2003. Synthesis and properties of superconductor $RuSr_2GdCu_2O_8$. *Physica C* 384: 137-142.

Williams, G. V. M., Jang. L. Y & Liu, R. S. 2002. Ru valence in $RuSr_2$ (Gd_{2-x} $Ce_x)Cu_2O_{10-\delta}$ as meansured by x-ray-obsorption near-edge spectroscopy. *Physical Review B* 65: 064508(1-5).

Yeoh,L.M., Ahmad, M., Abd-Shukor, R. 2008. Synthesis of $RuSr_2GdCu_2O_{8-\delta}$ superconductors by the sol–gel route. *Material letter.* 61:2451-2453.

Yeoh,L.M., Ahmad, M., Abd-Shukor, R. 2008. Superconductivity in Ru-based cuprate $Ru(Sr_{1.5}Ca_{0.5})PbCu_2O_{8-\delta}$ prepared by sol-gel route.

Yokosawa, T., Awana, V. P. S., Kimoto, K., Takayama-Muromachi, E., Maarit, K., Yamauchi, H. & Matsui, Y. 2004. Electron microscope studies of nano-domain structures in Ru-based magneto-superconductors: $RuSr_2$-$Gd_{1.5}Ce_{0.5}Cu_2O_{10-\delta}$ (Ru-1222) and $RuSr_2GdCu_2O_8$ (Ru-1212). *Ultramicroscopy* 98:283-295.

Zhigadlo, N.D., Odier, P., Marty, J.Ch., Marty, P. 2003. A. Sulpice, *Physica C.* 387: 347.

Chapter 7

CONCLUSION

Five types of wet chemical techniques have been successfully applied in the preparation of YBCO superconductors. In this study, annealing duration is fixed variables set for four hours while annealing temperature treatment is manipulated variable which is set at the temperature range 880-950 °C. Optimum annealing temperature treatment for these five specific techniques can be obtained from the result of the $T_{c\text{-zero}}$ of the samples prepared, known as the responding variable. Samples prepared by ASG and CT sol-gel techniques require low annealing temperature. High temperature treatment which will cause the evaporation or melting of some metal components and consequently destroy its stoichoimetry. For ASG-SSR, COP and COP-SSR techniques formation of superconductor is partially dependent on infiltration and diffusion process between the grain particle. High annealing temperature treatment is required to stabilized YBCO superconductor. However, the conditions and annealing temperature treatment has not been achieved optimums for this three techniques because the result are not getting any $T_{c\text{-zero}}$ of 90 K for YBCO samples. The duration of annealing treatment temperature should be extended more than four hours so that foreign phase can be eliminated because infiltration rates between the grains particle are slow for YBCO superconductor formation.

The orthorhombic structure of the Y-123 system was maintained throughout the range of percentage nano Ag added. The normal state resistance-temperature curve showed a metallic behavior for all the pure and nano Ag-added samples with $T_{c\text{-zero}}$ around 83–90 K, except for the 15% weight sample which showed semiconductor-like behavior at normal state with $T_{c\text{-zero}}$ 78 K. This is partly due to the change in the nature of the grain boundaries when the amount of nano Ag is increased to a certain limit. Nano Ag addition in YBCO prepared by ASG gel method does not suppress $T_{c-onset}$ (about 91–93 K) of all the samples studied. Nevertheless, Ag element behave as flux pinning in the YBCO samples and prepared a good conductivity path for copper pair electron between grain boundary, thus transport critical current density for samples improve. The transport critical current density, J_c increases with increase in Ag content. YBCO+10 % nano Ag shows the highest J_c value, 2.3A/cm^2 with T_{c-zoro} 84 K. These results shows that the present of nano Ag is useful in enhancing the transport current and flux pinning properties for practical applications such as in the fabrication of high temperature superconductor tapes and conductors.

The double doping superconductor $Y_{0.9}Ca_{0.1}Ba_{1.8}Sr_{0.2}Cu_3O_{7-\delta}$ was successfully synthesized through ASG-SSR with $T_{c-onset}$ close to 86 K and T_{c-zero} close to 80 K. From XRD pattern, the orthorhombic structure was maintained for this bulk superconductor with a = 3.887 Å, b = 3.829 Å, and c = 11.690 Å, which are not much different compared to value for the orthorhombic phase of undoped YBCO. That has shown that it is possible to obtain nearly single phase $Y_{0.9}Ca_{0.1}Ba_{1.8}Sr_{0.2}Cu_3O_{7-\delta}$ superconductor by this technique. Ca–Ba–Sr–Cu–O sol-gel powder and Y_2O_3 were used as starting material for this bulk superconductor preparation. The precursor Ca–Ba–Sr–Cu–O was prepared by conventional sol–gel method based on acetate–tartrate process. FTIR result for this precursor show that the apparent of hydroxyl group, –OH in the range 3600–3400 cm^{-1} are chemically reactive to Y_2O_3, therefore enhance the diffusion rate during solid-state reaction. The decrement in critical temperature, T_{c-zero} for this bulk superconductor compare to pure Y-123 is due to lattice distortion and deficient in oxygen content by the effect of Ca and Sr doping even in very small amount. Sol–gel method allows a homogeneous mixing of the metal precursors on an atomic scale, therefore provides an efficient means of synthesizing multicomponent ceramics in a very stoichiometry amounts.

The $RuSr_2GdCu_2O_{8-\delta}$ (Ru-1212) superconductor was successfully synthesized through AGS sol–gel method with $T_{c-onset}$ near 55 K and T_{c-zero} at 45 K. Optimization of synthesis and annealing conditions may further improve the superconducting property. We found that 1030 °C is the optimum annealing temperature for the Ru-1212 synthesized by the sol–gel route. Hence, the sol–gel route is an alternative way to prepare the Ru-1212 superconductor. This technique is more effective and only involves one step of calcinations reaction.

ASG sol-gel technique was also successfully applied to the preparation of $RuSr_2(Gd_{2-x}Ce_x)Cu_2O_{10-\delta}$ system. Consistent with past research which is prepared by solid-state reaction, this route resulted in the formation of single-phase superconducting Ru-1222 for x = 0.6, in a one-step process with $T_{c\text{-zero}}$ 40 K. Higher Ce content will destroy nature superconductivity of Ru-1222 system.

New phase formation of Pb containing ruthenium-based superconducting cuprate $Ru(Sr_{1.5}Ca_{0.5})PbCu_2O_{8-\delta}$ (Ru-1212 type) have been successfully synthesized through the sol–gel route. The optimum annealing temperature for the ruthenium-based cuprate superconductor was found to be 890 °C. The XRD result shows a single Ru-1212 type phase with tetragonal symmetry. This material shows $T_{c\text{-onset}}$ at 35 K and $T_{c\text{-zero}}$ at 20 K. Pb together with the partial substitution of Sr with Ca lowered the formation temperature of the Ru-1212 type phase.

PROPOSED SYUDY

Three projects have been successfully conducted in this research study. Overall, the results of this study was to achieve research objectives. However, there are several aspects of the study can be done to complete the understanding in the field of superconductors. Some suggestions are listed as follows:

1) The first research project involving the application of five wet chemical techniques in the preparation of YBCO superconductors. Further studies may be conducted on the YBCO samples prepared by ASG-SSR, COP-SSR and COP techniques that have not reached the optimum conditions. Period of heat treatment for samples can be extended more than 4 hours, so that foreign phase such as Y-211 and $BaCuO_2$ can be eliminated because infiltration rates between the grain particles to form orthorhombic structure, Y-123 are slow in order to get stable YBCO superconductor. The result of $T_{c\text{-zero}}$ according to the annealing temperature and duration are investigated. XRD pattern and change of lattice parameters a and c can be studied further. Distribution of internal structures can be observed by SEM and EDAX.

2) The second research project is the effect of nano size Ag addition on

YBCO superconductor. Addition of other transition metal elements such as nano size manganese (Mn) and chromium (Cr) which have nature of magnetic properties are used to replace the nano Ag on YBCO additional. Changes in transition temperature, T_c and transport critical current density, J_c value. The additional of weight percentage of metal elements of Mn and Cr and its impact on phase purity 123 can be investigated and studied.

3) In the system of $RuSr_{1.5}Ca_{0.5}PbCu_2O_{8-\delta}$ (RuPb-1212) with multivalensi elements such as Ce, and Pr can be use to replace the Ca element. The existence of these elements in a single or mixture valence and changes to the system of RuPb-1212 could be investigated.

INDEX

S

T